Crucible of Science

Crucible of Science

THE STORY OF THE CORI LABORATORY

JOHN H. EXTON

Oxford University Press is a department of the University of Oxford.
It furthers the University's objective of excellence in research, scholarship,
and education by publishing worldwide.

Oxford   New York
Auckland   Cape Town   Dar es Salaam   Hong Kong   Karachi
Kuala Lumpur   Madrid   Melbourne   Mexico City   Nairobi
New Delhi   Shanghai   Taipei   Toronto

With offices in
Argentina   Austria   Brazil   Chile   Czech Republic   France   Greece
Guatemala   Hungary   Italy   Japan   Poland   Portugal   Singapore
South Korea   Switzerland   Thailand   Turkey   Ukraine   Vietnam

Oxford is a registered trademark of Oxford University Press in the UK and certain other
countries.

Published in the United States of America by
Oxford University Press
198 Madison Avenue, New York, NY 10016

© Oxford University Press 2013

All rights reserved. No part of this publication may be reproduced, stored in a
retrieval system, or transmitted, in any form or by any means, without the prior
permission in writing of Oxford University Press, or as expressly permitted by law,
by license, or under terms agreed with the appropriate reproduction rights organization.
Inquiries concerning reproduction outside the scope of the above should be sent to the
Rights Department, Oxford University Press, at the address above.

You must not circulate this work in any other form
and you must impose this same condition on any acquirer.

CIP data is on file at the Library of Congress
ISBN 978-0-19-986107-1

Dedicated to Charles Rawlinson (Rollo) Park

Contents

Acknowledgments ix
Introduction xi

1. Carl and Gerty Cori: Early Life   1

2. Sidney Colowick: Their First Graduate Student   16

3. Herman Kalckar: The Great Dane   21

4. Severo Ochoa: Spanish Genius   29

5. Move to Enzymology and Work of Arda Green   39

6. Luis Leloir: One of Argentina's Greatest Scientists   45

7. Earl Sutherland: Master of Intuition   53

8. Coris' Move to the Department of Biological Chemistry: Award of Nobel Prizes and Career of Tom Cori   67

9. Sidney Velick: Modest Enzymologist   73

10. Victor Najjar: Pediatrician and Immunochemist   78

11. Edwin Krebs: Accidental Biochemist   82

12. Mildred Cohn: Against All Odds   91

13. Christian de Duve: Belgian with Savoir Faire 100

14. Arthur Kornberg: A Giant of Biochemistry 112

15. Hormone Effects on Muscle Carbohydrate Metabolism 123

16. Charles Park: Aristocratic Physiologist 127

17. Jane Harting Park: Enthusiast for Science 139

18. Gerty Cori's Work on Glycogen Structure and Glycogen Storage Diseases 145

19. Joseph Larner: Focus on Glycogen Synthase 149

20. Contributions of Barbara and David Brown 161

21. William Daughaday: All About Growth 165

22. Robert Crane: A Decade with Carl Cori 171

23. Alberto Sols: Spanish Enzymologist 176

24. Luis Glaser: The Complexity of Carbohydrates 181

25. Ernst Helmreich: Jovial Bavarian 189

26. Carl Frieden: Enzyme Kineticist 196

27. David Kipnis: Focus on Diabetes 201

28. William Danforth: Academic Leader 206

29. The Influence of the Coris on Washington University and Carl Cori's Research at Boston 210

30. The Heritage of the Coris 214

Index 221

Acknowledgments

Regrettably, many of the scientists who worked in the Cori laboratory have passed on and I have had to depend on written sources to reconstruct their lives and accomplishments. However, I am grateful to some of them who are still active and have great recollections of the laboratory. In this regard, I would like to thank Rollo and Janey Park, Joe Larner, Ernst Helmreich, Barbara (Bobbie) and David Brown, Luis Glaser, and Carl Frieden. I was also greatly helped by conversations with Maryda Colowick, Jennifer and Tilly Najjar, Tom Cori, Joel Hardman, and Murray Heimberg.

No work of this magnitude can be accomplished without the assistance of many people. Thus I would particularly like to thank Jeremy Lewis and Hallie Stebbins my editors at Oxford University Press, and Maureen Cirnitski for superb copyediting. I am particularly grateful to Phil Skroska, Archivist, Visual and Graphic Archives at the Bernard Becker Medical Library at Washington University. Other important people were James Thweatt and Christopher Ryland of the Special Collections, Eskind Biomedical Library, Dominic Doyle of the Medical Art Group; and Kevin Davis of the General Counsel, all of Vanderbilt University. Others who provided help at various stages were Douglas Atkins of the National Library of Medicine, Louis Hue and Monique Van de Maelen at the de Duve Institute, Brussels. An important contributor to the book was my Vanderbilt colleague Owen McGuinness, who provided much-needed word processing skills. Finally, but not least, I thank my wife Janet for her continuous support during the longer than expected birth of this book.

Introduction

This is the story of Carl and Gerty Cori who established a laboratory at Washington University in St. Louis, Missouri, when Carl was appointed chair of Pharmacology in 1931 and chair of Biological Chemistry in 1946. The Cori laboratory remains unique in its contributions to biomedical science. Not only did both the Coris win the Nobel Prize, but also six other Nobel Laureates emerged from their laboratory, a record unmatched by any other U.S. institution. The discoveries of the Coris and their associates led to great advances in knowledge of how the body controls the level of glucose in the blood and how insulin and other hormones affect this. They defined the mechanisms by which the hormones epinephrine and glucagon control the breakdown of glycogen, which is the storage form of glucose in the body. They studied the uptake of glucose from the gut and tracked what happened to it in the body. They identified many of the reactions involved in the processing of glucose and glycogen in the body and characterized many of the enzymes catalyzing these reactions. They also studied how glucose was taken up by muscle and how hormones influenced this. Thus, their investigations into carbohydrate metabolism were of great relevance to normal physiology and metabolic diseases such as diabetes.

The Coris were both born in Prague in 1896 and met at the same medical school there. After service in the Austrian Army during World War I, Carl finished his clinical studies and went to Vienna to marry Gerty. Because of Gerty's Jewish heritage and the general conditions of life in Europe at that time, they decided to emigrate to the United States in 1922 where they took positions at a cancer research institute in Buffalo. This is where they carried out their classic studies of the effects of adrenaline on glucose and glycogen metabolism that led to the formulation of the Cori cycle. The Cori cycle is a metabolic cycle whereby lactic acid released into the bloodstream by the breakdown of glycogen in muscle is reconverted to glycogen in the liver, and glucose produced by the breakdown of glycogen in the liver is returned to the muscle.

In 1931, Carl was surprised to receive an offer from Washington University in St. Louis to become chairman of Pharmacology. When they took up their appointments, the Coris resolved that real advances in their field of research would come from the use of isolated tissues and the purification and characterization of the enzymes involved in glucose and glycogen metabolism. This decision profoundly affected the course of their research, and their laboratory at Washington University was soon recognized to be at the forefront of biochemical research in the United States. Even before the Coris received the Nobel Prize in 1947, young scientists flocked to their laboratory from all over the United States and the world.

The Coris had rather formidable personalities and were rigorous and demanding mentors. They were intolerant of sloppy thinking and emphasized that any projects should be truly novel and any data should be reproducible and internally consistent. They insisted that data should be analyzed quantitatively where applicable and that results should not be interpreted in terms of any preconceived notions, but with due consideration of alternative explanations. In short, their laboratory provided a superb training in the scientific method.

Many of the scientists that trained in the Cori laboratory went on to make major discoveries and to win Nobel Prizes (Figure 1), and the careers of many of these are described in this book. Notable among them was Herman Kalckar, a Dane with an exceptionally creative mind who made wide-ranging discoveries and interacted with an astonishing number of Nobel Laureates. He made many important contributions to biochemistry. The Coris' first graduate student was Sidney Colowick, who had never taken a course in biochemistry and was qualified by the U.S. government as a shrimp inspector! He turned out to be a very quick learner and was described by Kalckar as having a brilliant analytical mind and being wise beyond his years. His experimental expertise was instrumental in the Coris' discoveries in St. Louis. A major finding in which he played a key role was to define the product of phosphorylase action on glycogen, which was, unexpectedly, glucose-1-phosphate. Colowick later achieved fame as coeditor of *Methods in Enzymology*, a series of volumes popularly known as the biochemist's bible. He interacted with many other scientists in the laboratory, but in particular with Earl Sutherland. Sutherland was gifted with great intuition, and his work in St. Louis on the hormonal regulation of phosphorylase, the enzyme responsible for glycogen breakdown, took a leap forward when he showed that the enzyme was activated by the incorporation of phosphate (phosphorylation) and inactivated by the release of this phosphate (dephosphorylation), reactions involving a protein kinase and a protein phosphatase, respectively. Furthermore, he showed that epinephrine and glucagon promoted phosphorylation of the enzyme. His subsequent finding that the action of the hormones involved the formation of a heat-stable factor led to the discovery of cyclic AMP. This nucleotide is now recognized to be the intracellular messenger for many hormones, neurotransmitters, and drugs. For his discovery of cyclic AMP, Sutherland was the sole recipient of the Nobel Prize in Physiology or Medicine in 1971. In his early work in the Cori laboratory, Sutherland studied glucagon, which is counterregulatory to insulin in controlling the blood glucose, and in this he was joined by Christian de Duve, a fellow from Belgium who would later win the Nobel Prize in 1974 for the discovery of lysosomes, which are intracellular organelles involved in the digestion of cellular components. De Duve further showed that reduced activities of certain lysosomal enzymes were responsible for several genetic diseases.

In 1947, the Coris received the Nobel Prize for their work on glycogen metabolism and shared it with the Argentinian scientist Bernardo Houssay, who studied the effects of anterior pituitary hormones on carbohydrate metabolism. The excitement and joy of the occasion was clouded by the devastating news that Gerty was suffering from an incurable form of anemia. She was the first American woman to receive the Prize and the third from anywhere in the world Despite the physical demands of the disease, she persevered in her research and made significant contributions on glycogen structure and glycogen storage diseases until her valiant struggle ended in 1957.

The brilliant Spanish scientist Severo Ochoa spent time in the Cori laboratory where he studied the metabolism of fructose. Ochoa was a genius who later made major contributions in two different areas. The first contribution was a study of the enzymes of the citric acid cycle, a cyclic series of enzymes that generate the bulk of the energy (ATP) utilized in cells. The second was his discovery of the enzyme polynucleotide phosphorylase. Although he was disappointed to realize that this enzyme was not involved in RNA synthesis, it proved to be extremely useful in deciphering the genetic code. This is because it could make polynucleotides that could be matched to specific amino acids in the process of protein synthesis—the point being that the genetic code could be broken by changing the base composition of the synthetic polynucleotides to see which amino acids were incorporated into the resulting polypeptides. For his contribution to the breaking of the genetic code, Ochoa received the Nobel Prize in 1959 together with Arthur Kornberg, who also worked in the Cori department.

Three other Nobel Laureates spent time in the Cori laboratory. One of these was Luis Leloir from Argentina who worked on the enzyme that carries out the first step in the citric acid cycle. On his return to Buenos Aires, he studied the synthesis of the disaccharide sugar lactose and found that a cofactor was required. This was shown to be an unusual compound of glucose and the nucleotide UDP (UDP-glucose). Subsequent work showed that combinations of other nucleotides and sugars were involved in the synthesis of sucrose, glycogen, starch, and a vast array of complex carbohydrates that are components of cell walls and are involved in many cellular processes. For his classic studies, he received the Nobel Prize in Chemistry in 1970.

Edwin Krebs spent three years with the Coris studying phosphorylase and one of the glycolytic enzymes involved in the breakdown of glucose. On leaving for a position at the University of Washington, Krebs continued his work on the control of phosphorylase activity, expanding the finding of Sutherland that the regulation of phosphorylase activity was due to the enzymatic addition and removal of phosphate. Working with his close collaborator Edmond Fischer, he found that cyclic AMP stimulated the activity of phosphorylase kinase, the enzyme that phosphorylates and activates phosphorylase. However, it took several years to show that cyclic AMP did not act directly on phosphorylase kinase, but on a separate contaminating protein, kinase, which was named cyclic AMP-dependent protein kinase. This kinase was found to act on a variety of cellular proteins altering their functions. The concept that enzymes and other proteins could be controlled by phosphorylation and dephosphorylation was found to be so widespread that Krebs and Fischer received the Nobel Prize in 1992.

One of the real stars who spent time in the Cori laboratory was Arthur Kornberg, arguably the greatest biochemist of his era. Remarkably, he received no formal training in biochemistry apart from a course in medical school. In St. Louis, his initial project was not productive, and he returned to the National Institutes of Health from which he was on leave. He was then recruited to Washington University as chairman of microbiology where he undertook the formidable project of determining how DNA was synthesized. After very arduous and time-consuming work, this resulted in the purification and characterization of DNA polymerase, the enzyme involved. This discovery of the mechanism of DNA replication was an outstanding accomplishment because of the fundamental importance of DNA as the carrier of genetic information. He received the Nobel Prize for this in 1959. The Coris were unhappy to see him move to Stanford where, after much trial and error, he was able to get DNA polymerase to assemble a 5,000 nucleotide chain which had the same genetic activity as DNA from a natural virus. This amazing feat was dubbed by the local press as the "creation of life in a test tube." His work on DNA replication underlies the massive program of sequencing and cloning the human genome and that of many other species, the manipulation of genes to produce altered life organisms, and the use of the polymerase chain reaction for the cloning of DNA from vanishingly small samples. It has resulted in a revolution that has transformed biology and led to the establishment of innumerable biotechnology companies using recombinant technology to produce large amounts of scarce proteins of medical and other importance.

Other investigators who received training in the Cori laboratory included a large number of outstanding scientists whose work focused on the effects of insulin, growth hormone, and cortisol on glucose uptake in muscle from normal and diabetic rats. Apart from explaining many of the metabolic changes in human diabetes, the researches showed that insulin promoted the uptake of glucose in muscle by stimulating its transport across the cell membrane. Many of these investigators became leaders in the areas of insulin and diabetes research. Their individual contributions are described in the various chapters of this book, but it is noteworthy that almost all of them rose to play significant leadership roles in academic medicine in the United States and overseas and to carry on vigorous research programs. In turn, they trained notable scientists, including Nobel Laureates.

Following Gerty's death in 1957, Carl was devastated and emotionally drained, but his spirits were lifted by a happy second marriage in 1960. He and his new wife, Anne Fitzgerald-Jones, shared many interests, and Carl's wit and grace flourished in this period of his life. He retired from the chairmanship of the Department of Biological Chemistry in 1966 when he reached the age of 70. The couple moved to Boston where he was appointed visiting professor of biological chemistry at Harvard with a laboratory at the Massachusetts General Hospital. He remained active in research until his death in 1984, with his area of interest being genetic studies of glucose-6-phosphatase. This work was carried out in collaboration with the noted geneticist Salomé Glüecksohn-Waelsh of the Einstein College of Medicine and revealed a novel finding, namely that mammalian enzymes could be regulated not only by genes specifying their structure, but also by genes that regulate their expression.

The Coris were very accepting of fellows from all over the world, and they also accepted women at a time when they had difficulty in finding appropriate positions. Gerty herself had encountered gender discrimination when she did not initially receive a faculty position at Washington University and was paid a much lower salary than Carl was. The first woman to join the group was Arda Green. She had trained with the eminent protein chemist Edwin Cohn and had an uncanny gift for purifying proteins. She worked with Gerty to purify and crystallize phosphorylase, and, after she left to work at the Cleveland Clinic, she purified proteins involved in the regulation of blood pressure. Another brilliant female investigator who worked in the Cori laboratory was Mildred Cohn. She had struggled to be accepted as a graduate student by the Nobel Laureate Harold Urey and, despite her tremendous gifts as a researcher, did not achieve faculty status until after twenty-one years as a research associate! She transformed research in the Cori laboratory by bringing to it the use of stable isotopes and the techniques of electron spin and nuclear magnetic resonance to define the mechanisms of enzymatic reactions involved in carbohydrate metabolism. Two other women who worked in the Cori department were Barbara Illingworth and Jane Harting. Barbara worked with Gerty Cori on glycogen structure and glycogen storage disease while Jane worked with Sidney Velick on the glycolytic enzyme glyceraldehyde-3-phosphate dehydrogenase.

It is interesting to speculate on what made the Cori laboratory so successful and attractive to so many brilliant scientists. One element was that they espoused the "new" biochemistry, which explored the reactions that were involved in metabolism, with a focus on the enzymes involved. Prior to this, biochemical research in the United States tended to be focused on nutrition, especially amino acids and vitamins. Another factor that directed people to the Coris was that the prewar powerhouses of biochemical research in Germany had been devastated by World War II and by the departure of many brilliant Jewish scientists, and the type of research previously conducted by these laboratories was now being performed by the Coris. As their reputation developed and the exciting nature of the research that emerged from their laboratory became widely known, their department became a "mecca" for many young scientists.

Concerning the success of the laboratory, a key element was the scientific rigor that the Coris demanded of their associates, with attention to all elements of the research process. There was also a high degree of interaction between members of the laboratory, and this often resulted in a synergy when two or more associates worked on a project. There was also a requirement that the associates present their data to the group, even in the intimidating presence of the Coris. All this rigor meant that the Coris attracted only able students and fellows, and the high quality of their associates almost ensured that the emerging research would be of the highest standard. They were unconcerned with issues of gender, religion, and national origin and, surprisingly, did not require prior training in biochemistry—it was the quality of the mind that mattered. With their intellectual power and demanding standards, they created a crucible out of which only the finest research and scientists emerged.

Crucible of Science

1

Carl and Gerty Cori

Early Life

Carl Ferdinand Cori was born to Carl I. Cori and Martha Lippich on December 5, 1896, in Prague, which was then part of the Austro-Hungarian Empire under the long-lived emperor Franz Josef. He was the middle child of three, with an older sister Elisabeth and a younger sister Margarete. He was born into a family of academics. His mother was born in Graz in 1870, and his father was born in Brüx, Bohemia, in 1865. Carl's maternal grandfather was Ferdinand Lippich, a professor of theoretical physics at the German University of Prague (Carl Ferdinand University) who developed the polarimeter as a precision instrument for identifying and measuring chemical compounds; an uncle, Friedrich Lippich, was a professor of chemistry there. Carl's father studied medicine and natural sciences at the German University of Prague, where he obtained an MD, and at the University of Leipzig, where he got a PhD in zoology. Upon graduation, he became a lecturer at the Zoological Institute in Prague, and in 1898, he was appointed professor and director of the Marine Biological Station in Trieste.

The family moved to Trieste when Carl was two years old. He became fascinated with the various forms of marine life at the station, especially *Spirographis*, a marine worm that would withdraw its feathery tentacles into its tube when touched, a torpedo fish that generated electric shocks, an octopus that changed colors, and *Holothuria*, which would lose its intestinal tract when disturbed, but then regenerate it. He especially enjoyed the field trips in the station's launch *Adria* when the contents of the dredge were emptied into a large canvas bag. His father could draw beautifully, and his lectures to the students included the history of the local Roman and Christian cultures as well as marine biology and oceanography. The Austrian zoologist Karl von Frisch, who won the Nobel Prize in 1973 for deciphering the honey bee dance, recalled in his memoirs the enjoyment of field trips on the *Adria*. Trieste was the sole Austrian port at the time Carl was there, and it provided the Austro-Hungarian Empire a gateway to the wider world and a teeming center of international commerce. It was a colorful city with narrow and chaotic streets and a mixed population of Italians, Greeks, Slovenes, Germans, Jews, and others. Its views of the Alps, interesting hinterland, and diverse languages had a profound effect on the young schoolboy. Trieste also provided a host of other biological experiences, including exploring the local caves and the Dalmatian coast for beetles and other insects and collecting lizards from islands in the Adriatic Sea.

Trieste was a wild town with its street urchins and colorful life along the quays where sailors roamed in search of girls and where the walls were embellished with much sexually oriented graffiti. Carl's "education" in this tolerant city was balanced by vacations in the Tyrolean Alps where he went on mountaineering trips and where his maternal grandfather, Ferdinand Lippich, exposed him to chamber music. Ferdinand was a gentle person with a unique sense of humor and a great repertoire of verses and stories.

Many scientists visited the Cori family in Trieste, and one encouraged Carl's interest in insects. Carl made a special study of the blind beetles that inhabited the limestone caves above Trieste, and to do this, he joined a spelunker's club called "Hades." The group would meet in a country pub and ply one of the local farmers with wine. After a bottle or two, the farmer could then be persuaded to take them to a deep hole in the vicinity where the group collected the beetles. By means of wire rope ladders, they made the dangerous descent of one hundred feet into the unknown. The caves had extensive galleries, and the threat of cave-ins and subterranean rivers made collecting the beetles a frightening experience. However, Carl described the explorations as providing much pleasure and inspiration.

As was traditional at that time, Carl entered a gymnasium where Latin and Greek were emphasized and the sciences virtually ignored. He stated that he was an indifferent student and felt that most of his education came from his experiences at the Marine Biological Station and the field trips in the Adriatic. He felt that his contact in school with many different language groups was a humanizing experience since it gave him some immunity against racist propaganda. Furthermore, the history of Austria was not one that inspired patriotic propaganda, with many battles lost, for example, Austerlitz, at which Napoleon won his most brilliant victory. In his autobiography, Carl states that he passed the strict examination that entitled him to enter one of the faculties of philosophy, jurisprudence, medicine, or philosophy, but felt like Goethe's Faust, who studied all the four disciplines but felt no wiser than before. He was not always a serious child but was given to inflict practical jokes on his family, such as turning the water pink by sprinkling potassium permanganate in the chamber pot used by a hypochondriac maiden aunt and disturbing a picnic party by releasing yellow jacket wasps. He also hid silkworms in the family parlor in the hope that they would hatch during an important party. They did, to the consternation of his mother, who was concerned about the housekeeping. One of his duties was to dust his father's library, which gave him the opportunity to explore the medical texts. These texts included the *Psychopathia Sexualis* by Richard Krafft-Ebing, in which he discovered some exciting aspects of human behavior. He recounted that, as a teenager on a trip to Venice conducted by an old professor, he managed to fall into the Grand Canal. The professor was fearful that Carl would contract pneumonia and ordered him to bed. In order to warm Carl up, he ordered his buxom maid to undress and get into bed with him. On reading that story, I have often thought of falling into a Venetian canal, but I doubt that the outcome would be so pleasant!

In the summer of 1914, when Carl had finished gymnasium, he was allowed to join his father's expedition to study the ecology of islands off the coast of Dalmatia, a part of Croatia. He was given the task of collecting insects, especially ants. He published a report on his findings and gave the collection to the Natural History Museum in Vienna. World War I had broken out during the expedition, but they had not heard about it until they put into Spalato, which is now called Split. Few had expected the assassination of Archduke Franz Ferdinand by a Serb in Sarajevo to result in war. However, the archduke was the great-nephew of Franz Josef and heir apparent to the Austro-Hungarian throne.

Medical Studies in Prague and Service in the Army

When the war broke out, Carl decided to become a medical student and entered the German University of Prague. The war seemed remote, and his first-year studies included chemistry, physics, and biology, but because of the war, his second-year course was accelerated to include anatomy, biochemistry, physiology, and pharmacology. While at the medical school, Carl was captivated by a fellow student, Gerty Theresa Radnitz, who would become his wife. She was a talkative extrovert, whereas he was quiet and reserved. They loved to take ski trips together and hikes in the countryside until Carl was drafted into the Austrian Army. He first served with ski troops and was almost killed. He was then transferred to the Sanitary Corps as a lieutenant and was assigned to a bacteriological laboratory where he contracted typhoid fever despite receiving the typhoid vaccine.

To his surprise and delight, he was transferred to Trieste to set up a laboratory in the Marine Biological Station. He didn't stay long because a fellow officer reported him for being late returning from furlough, and, as a consequence, he was sent to the front lines. However, he never made it to the front because of an automobile accident in which he was able to stop the bleeding of a passenger from a cut temporal artery. In his autobiography, Carl recounts a later episode when a professor asked a medical student what he would have done in such a case. The student replied that he would put a tourniquet around the neck and was promptly flunked! Carl spent the last year of the war at a hospital on the Piave River in Italy, close to the front lines. This hospital was staffed by physicians from the University of Budapest who trained him in clinical medicine and pathology. He recounted that one of his duties was to look after the local civilians because of his proficiency in Italian. On entering the homes of the poor farmers, he was invariably offered a glass of wine, but encountered fleas leaping from the patients' bed clothes when he tried to examine them. There were outbreaks of smallpox, tuberculosis, malaria, and typhoid fever, and pellagra and scurvy were rife. This was due in part to poor vaccination, but mainly to ignorance of the population. At the end of the war, the situation became chaotic as the army retreated from the Piave River. Carl escaped by riding on the roof of a freight car, living through an attack by a low-flying plane. The barbarism of war and the ravages inflicted on civilians left a profound impression on Carl. He noted that the influenza epidemic—with its high mortality rate among the poorly nourished soldiers and civilians and the inability to be of any help—came as a great shock to him.

Carl returned to Prague in 1918 and finished his clinical training there, graduating with an MD in 1920. In August of the same year, he married Gerty in Vienna, where they had both gone for postdoctoral work. At that time, World War I was over and Prague had become the capital of the newly formed Czechoslovakia. Carl and Gerty greatly admired Thomas Masaryk, the president of the new republic, but recognized that the new government had more pressing needs than to promote research. Consequently, they felt it was best to remain in Vienna. There, Carl engaged in laboratory work in the university's Internal Medicine Clinic in the mornings and in the Pharmacological Institute in the afternoons. His remuneration at the clinic was one free meal a day! His work at the Institute was not his first experience of research; in 1919, he had worked as a summer student in the laboratory of Albert Szent-György in the Pharmacological Institute of Dr. Geza Mansfeld. Szent-György would later win the Nobel Prize in 1937 for his work on biological oxidations and vitamin C. The laboratory was located in the newly founded Elisabeth University in Pozsony, an old Hungarian town; but as a result of the Treaty of Versailles, the town was given to Czechoslovakia and renamed Bratislava. To prevent the Czechs from appropriating the equipment, Cori and Szent-György dressed as workmen and in the middle of the night removed all of the equipment through the closely guarded gates of the campus and transported it across the Danube.

Carl became disillusioned with clinical medicine, but he was excited by research. His preceptor at the clinic in Vienna was brilliant, but unethical and anti-Semitic. Carl was shocked by some of the practices in the laboratory and came to doubt whether he should continue in the practice of medicine and even if he and Gerty should stay in Europe. However, life was not hard, and they enjoyed the cultural richness of Vienna with its wonderful art museums—especially the Albertina, with its collection of Albrecht Dürer prints, and the Belvedere, which featured the brilliant Austrian artists Gustav Klimt and Oskar Kokoschka. However, the jewel in the crown was the Kunsthistorisches Museum, notable for its great collection of Pieter Breugel the elder, Rubens, Dürer, Velasquez, and its wonderful collection of Italian Renaissance paintings, including works by Leonardo, Titian, Tintoretto, Raphael, Caravaggio, and Bellini. Carl and Gerty were also intrigued by the adjacent Museum of Natural History, with its prehistoric section containing objects from the salt mines of Hallstatt and the famous Venus of Willendorf statuette. This was the museum to which Carl donated his collection of ants and other insects. Although entrance to the museums was mainly free, attendance at concerts and the opera were beyond their means.

Compared with many details of Carl Cori's early life, information on Gerty's early days is more limited. She was born Gerty Theresa Radnitz on August 8, 1896, in Prague to Martha Neustadt and Otto Radnitz. Her father was a chemist who devised a method for purifying sugar and later became the successful director of a sugar refinery in Bohemia, now in the Czech Republic. Gerty's mother was a cultured woman who was a friend of the novelist Franz Kafka. One of her uncles was professor of pediatrics at the University of Prague. She was taught at home until she was ten and then at a lyceum for girls, graduating in 1912. She wanted to study chemistry at the university, but did not meet the requirements. So she went to a college preparatory school (Tetschen Real Gymnasium) for two years, and, after passing the entrance examination, which included literature, history, Latin, mathematics, physics, and chemistry, she enrolled at the same medical school as Carl in 1914, graduating in 1920, the same year as Carl. In his autobiography, Carl described her as a young woman with charm, vitality, intelligence, a sense of humor, and a love of the outdoors. While in medical school, they studied together and went on excursions to the countryside. As was typical of that time, Gerty was one of only a few female medical students.

In 1920, Gerty did her postdoctoral work in pediatrics at a children's hospital (Karolinen Kinderspital) in Vienna. Europe was still recovering from the deprivations of the war with limited food, but Gerty refused dietary supplements at the hospital, feeling that the patients needed them more. In the year following their marriage, Gerty developed xerophthalamia, a form of blindness. At that time, it was not recognized that this eye condition was due to vitamin A deficiency. However, her condition was cured by the better diet she received when she returned home to Prague. Her research at the hospital involved studying thyroid action in a patient with myxedema (thyroid deficiency) and in thyroidectomized rabbits. The idea was to see the influence of the thyroid on temperature regulation. She also did research on patients with certain hematological disorders. Since Gerty was Jewish, the chances of her obtaining a university position seemed remote, and this was an important factor in the Coris' decision to move to America.

Move to the United States

In 1922, Dr. Gaylord, the director of the State Institute for the Study of Malignant Disease in Buffalo, New York, was touring Europe in search of a biochemist for the Institute. He felt that an Austrian scientist would be more acceptable in the United States than a German and had come to Vienna to seek the advice of several senior investigators, including H. H. Meyer, the former director of the Pharmacological Institute where Carl worked. Meyer was notable for the development of the Meyer-Overton theory of anesthesia, which related the potency of anesthetics to their lipid solubility. Carl didn't feel he had made a positive impression on Dr. Gaylord and felt that the director wasn't interested in him. This was because he only had two interviews, one of which was in a car when Gaylord talked mostly to another doctor, and another while he was doing chemical tests on a patient's urine. He resigned himself to not getting the job and moved to the University of Graz to work with the pharmacologist Otto Loewi. Loewi had just made a major discovery. This was the chemical basis of nerve transmission, which he showed by demonstrating that stimulation of the vagus nerve to the heart released a chemical that was later shown by Sir Henry Dale to be acetylcholine. Loewi shared the 1936 Nobel Prize with Dale for these discoveries. Loewi was a stimulating teacher and had a wide knowledge of pharmacology, physiology, and biochemistry, and during his stay there, Carl formulated some ideas that he would later test in the United States. Although Carl was greatly impressed by Loewi's enthusiasm, originality, and wide knowledge of biomedical science, he didn't stay long in his laboratory because Loewi was Jewish, and those who couldn't demonstrate a pure Aryan descent were subject to dismissal by the university. Carl himself was not Jewish, but he still had to prove it and felt the whole process to be highly repugnant.

The general conditions of life in Europe at this time—together with Gerty's Jewish heritage and the premonition that forces were at work that would lead to another war—motivated them more strongly to move to the United States. However, the job at Dr. Gaylord's institute in Buffalo was still uncertain, and so they applied to the Dutch government for service among the natives of Java. In his autobiography, Carl wondered what their future would have been had their application been accepted!

Carl finally obtained the job at the Institute in Buffalo and moved there in 1922, and Gerty followed him six months later. At that time, it was considered improper for a husband and wife to be separated for any length of time, but Gerty felt a dedication to her job in Vienna. In Buffalo, Carl's job was to run the routine laboratory for the hospital and perform special tests as needed. One of the instruments he used was Duboscq's colorimeter, which compared the color of a test substance with that of a standard. He was free to spend the rest of his time in research of his choosing, which nobody was interested in or seemed to care about. The director, Dr. Gaylord, was ill and died two years after Carl had arrived. Carl quotes his successor as saying: "Gentlemen, it behooves us to find the cause or cure of cancer, and it's got to be intravenous." The blessing was that Carl was left to his own devices. Despite its erratic beginnings, the Institute is now the Roswell Park Memorial Institute and has a stellar reputation.

Upon her arrival at the Institute, Gerty was assigned to the Pathology Department, working principally on microscopic diagnoses. She encountered difficulties with Dr. Gaylord, who firmly believed in the parasitic basic of cancer. The problem arose when, in contrast to a previous report, she was unable to identify *Ameba histolytica* in the stools of cancer patients. This provoked the director to threaten to fire her unless she stayed in her room and quit working with Carl! When the storm calmed down, she worked on the effects of thyroid hormone on the multiplication of the unicellular organism *Paramecium caudatum* and later studied the effects of X-rays on skin and body organs.

Studies of Glucose Metabolism

There was initially great opposition to Carl and Gerty working together because of the prevailing antipathy to family members working in the same department or institution. However, this crisis subsided, and they were also excused from hospital service duties. One of their first great contributions came when they began collaborating on a study of the regulation of carbohydrate metabolism. They had been impressed by the work of Otto Warburg who, working at the Kaiser Wilhelm Institute (now Max Planck Institute) in Berlin-Dahlem, had discovered yellow enzymes critically involved in cell respiration. These are now called dehydrogenases and require the yellow coenzymes nicotinamide adenine dinucleotide (NAD) and flavin adenine dinucleotide (FAD). These discoveries earned Warburg the Nobel Prize in 1931. Warburg had had a very notable career. He was born in Freiburg in 1883, and his father, Emil Warburg, was an eminent physicist. He studied under the great organic chemist Emil Fischer, who won the Nobel Prize 1902 in Chemistry for his work on sugar and purine synthesis and who is often referred to as the "Father of Biochemistry." Warburg was an accomplished equestrian and served in the Prussian Horse Guards during World War I. Although Warburg was Jewish, the Nazis didn't remove him from his position, apparently because he was such a towering figure in German science and because his research involved cancer and Hitler had a deathly fear of this disease.

In addition to his pioneering studies of respiration, Warburg had made another significant observation, namely, that tumors studied in vitro produced large amounts of lactic acid. Although the basis for this effect is still not understood, it intrigued the Coris, who began a study of the phenomenon in vivo using mice with mammary tumors, chickens transplanted with Rous sarcoma tumor cells and a patient with a sarcoma. In all cases, the results showed that more glucose was converted to lactic acid in the tumor tissue compared to normal tissue. The Coris expanded this work to a study of the effects of insulin and epinephrine (adrenaline) on the levels of glucose and lactic acid in the blood and on the glycogen content of the liver and muscles of experimental animals.

Glycogen was discovered in the nineteenth century by the great French physiologist Claude Bernard and was later shown to be a complex polymer of glucose. It was also known that liver glycogen could break down to release glucose and that this was one of the means by which the level of glucose in the blood could be kept constant. It was already known that insulin reduced the glucose level and that epinephrine raised it, but the mechanisms underlying these changes had not been defined.

In his first experiments in Buffalo, Carl showed that epinephrine increased blood lactic acid, but insulin increased it only when doses sufficient to induce hypoglycemic convulsions were used, that is, when the large fall in blood sugar triggered the release of epinephrine from the adrenal glands. Carl used rabbits and cats in these experiments and was very conscious of the need to keep the animals calm. Thus he used rabbits with a "calm" disposition, and the cats were held in an assistant's lap and purring was taken as a sign that they were relaxed during the procedures. Likewise, rabbits were made as comfortable as possible and muscular movements minimized. He also went to great lengths to find the best method for measuring lactic acid. Large numbers of animals were used to ensure that the results were reproducible, and measurements were made at different times.

In further work to define factors regulating the blood glucose level, Carl studied the absorption of different sugars from the intestinal tract of rats. He found that the rate of absorption of each sugar was constant, but differed among the different sugars. Somewhat surprisingly, the rate of absorption was independent of the amount of sugar fed or its concentration in the intestine. When one examines these experiments, one is struck again by the great attention to accurate quantitation and experimental detail. For example, the sugars were administered by stomach tube, and any sugar adhering to the outside of the withdrawn catheter was checked, as was the amount of sugar that could be recovered from the intestine immediately after administration. Carl even tested to see if there was any sugar in the intestine prior to the experiments—even though the animals had fasted for forty-eight hours—and made sure that the metabolism of sugars by the bacteria in the gut was negligible.

To find out where the glucose was going in the body and where insulin was acting, Carl carried out some simple, but informative experiments with Gerty. These involved measuring the glucose levels in the arteries and veins of experimental animals (rabbits) and also in the vessels of normal and diabetic human subjects before and after treatment with insulin. In the case of the rabbits, samples were taken from the femoral (leg) vessels, but with humans, blood was taken from the finger tip (arterial) and from an antecubital (elbow) vein. The results showed unequivocally that in both species, insulin caused glucose to be taken up by muscle. In their paper describing their results, Carl and Gerty were very critical of the work of J. J. R. Macleod, who had been unable to detect a consistent effect of insulin, and he, in return, criticized their findings. Macleod had won the Nobel Prize in 1923, together with Frederick Banting, for the discovery of insulin. This decision by the Nobel Committee was reached with difficulty and was roundly criticized because the discovery had been made by Banting and Charles Best in

Toronto while Macleod was mostly out of the country on an extended visit to Scotland. The relations between Banting and Macleod became extremely bitter, and Banting initially refused to accept the Prize, but he gave in and announced that he would give half the money to Best. In an unusual event, the Nobel Foundation recognized in 1962 that Macleod had been wrongly given his share of the Prize.

Discovery of the Cori Cycle

The greatest contribution at this stage in their careers came when the Coris studied the influence of epinephrine on carbohydrate metabolism. In their first paper on this topic, they engaged in a thought experiment, namely, how does a fasting animal show only a minimal loss of glycogen from the body even though glucose is being used as a fuel? With remarkable prescience, they suggested that glucose was formed from noncarbohydrate sources in the liver and transported to the muscles by way of the bloodstream. In this way, glycogen levels in the liver and muscles could be maintained. They termed the process of glucose formation from noncarbohydrate sources gluconeogenesis. In their actual experiments, they observed that the injection of epinephrine caused an increase in blood glucose accompanied by a loss of muscle glycogen but an increase of liver glycogen. The changes in glycogen were puzzling, because it was known from earlier work by Gowland Hopkins at Cambridge and Otto Meyerhof at Berlin-Dahlem that the breakdown of muscle glycogen released lactic acid, but not glucose, and therefore this could not per se contribute to the increase in blood glucose. Furthermore, it was generally accepted that epinephrine mobilized liver glycogen to produce glucose. However, they found an increase and not a decrease in liver glycogen. In other words, the changes in glycogen in both muscle and liver could not explain the increase in glucose. With great intuition, the Coris expanded upon their thought experiment and came up with a theory to explain the effects of epinephrine. They posited that muscle glycogen was an indirect source of blood glucose because its breakdown induced by epinephrine produced lactic acid, which was carried to the liver by the bloodstream to be converted to glycogen there. This could be converted subsequently to glucose for use by the muscles.

To support their hypothesis, the Coris carried out a series of careful quantitative studies of carbohydrate balance in rats in the post-absorptive state. This entailed measurements of glucose uptake by the gut and simultaneous measurements of glucose oxidation and glycogen changes in the liver and muscle of control rats and those injected with epinephrine or insulin. In agreement with their earlier findings, the Coris observed that epinephrine decreased the deposition of the administered glucose as muscle glycogen, but increased its storage as liver glycogen. As expected based on results from previous studies, epinephrine raised the blood levels of glucose and lactic acid. Measurements in arterial and venous blood indicated that the lactic acid came from peripheral tissues, i.e., muscles, a finding that clearly supported their hypothesis. Insulin was observed to decrease liver glycogen, but this result was complicated by the fact that their insulin preparations were contaminated with another hormone (glucagon), which mobilized liver glycogen. The presence of this contaminant was unknown to the Coris, although it had been detected soon after the discovery of insulin.

In all of their experiments, they made extensive measurements and elaborate calculations to quantify the disposition of glucose, glycogen, and lactic acid. They examined and rejected alternative explanations for their data, mainly for quantitative reasons. They calculated, for example, that any mobilization of liver glycogen could not account for the increase in blood glucose induced by epinephrine and deduced correctly that the major factor was a decrease in the utilization of glucose by muscle. They also measured O_2 consumption and the respiratory quotient (the ratio of CO_2 produced to O_2 consumed) and also urinary N to get a measure of any changes in the oxidation of body fat and protein produced by epinephrine. The only significant changes were small increases in CO_2 consumption and calorie production. These were attributable to an increase in fat oxidation, which would not yield glucose. They also excluded the possibility that the action of epinephrine on glycogen breakdown and lactic acid production in muscle was due to vasoconstriction, that is, reduced supply of oxygen. More direct support came from studies that showed that the administration of lactic acid to rats led to the deposition of glycogen in the liver, with the natural d-isomer being more effective than the l-isomer.

At this time (1929), the Coris were sufficiently sure of their hypothesis that they included a diagram of it in their paper in the *Journal of Biological Chemistry*, although they had described the cycle in an earlier paper. The diagram shows that the breakdown of muscle glycogen induced by epinephrine releases lactic acid, which is then conveyed via the bloodstream to the liver to be converted to glycogen. This series of metabolic events became known as the Cori cycle. It was formulated on the basis of the actions of epinephrine and helps explain how activation of the sympathoadrenal system can counteract insulin-induced hypoglycemia. The cycle is also important in exercise for two reasons. Strenuously exercising muscles produce lactic acid, and the cycle provides a means of removing this from the blood, thus minimizing the problem of acidosis. Furthermore, the conversion of lactic acid to glucose and glycogen in the liver and the subsequent release of glucose, helps maintain the supply of fuel to the contracting muscles.

The Coris' findings were not universally accepted. In fact, Walter B. Cannon, a professor of physiology at Harvard and the most famous American physiologist at that time, wrote scathingly about their findings in a lengthy footnote in a review article. Cannon's work on the effects of stress on the body, epitomized in the so-called "fight or flight" response, had led to the hypothesis that epinephrine released from the adrenal glands during stress raised the blood glucose by breaking down liver glycogen. This was contrary to the data of the Coris, who found that epinephrine caused an increase rather than a decrease of liver glycogen. They were not intimidated by Cannon and wrote a spirited reply to him in a long footnote in one of their papers. As is often the case, both parties were correct, because epinephrine does cause an initial decrease in liver glycogen, which is later supplanted by an increase due to its synthesis from lactic acid. In fact, in later experiments where the Coris examined early changes in liver glycogen induced by epinephrine in rats, they found a transient decrease at fifteen minutes.

When one analyzes these early publications of the Coris, certain features emerge. These include their detailed knowledge of the literature, their selection of important questions, their logical approach to their experiments, their reliance on quantitation, and their rigorous analysis of their findings. In the preambles to their papers, they were often critical of experiments conducted by other investigators that were poorly controlled, did not use enough animals, or used sloppy methods. The Coris, on the other hand, always used sufficient animals and often more than one species. They paid meticulous attention to the accuracy and specificity of their analytical procedures and recorded the experimental conditions in great detail. For example, in studies of muscle, they were careful to make sure that the animals were not agitated and moving and, when studying the effects of insulin, they were careful to segregate out those animals that exhibited convulsions due to the hypoglycemic effects of insulin from those that did not; and in experiments involving the injection of epinephrine, they were careful to check that a physiological dose was used. Their rigorous use of quantitation and their concern about the reproducibility of their findings were hallmarks of their work, meaning that any observations could be relied upon since they were not just qualitative, but were also supported by many measurements. Thus certain explanations for the data could be ruled out simply on the basis that the measurements didn't add up. This meant that they could proceed to the next question without wasting time on hypotheses that did not have a logical basis. They were true pioneers in that very little was known about the regulation of carbohydrate metabolism when they began their work.

When one looks at the eighty publications that emerged from their work at the Institute from 1925 to 1931, about fifty were coauthored by Gerty, and she had eleven of her own. There are few coauthors besides the Coris, and the amount of research accomplished is staggering. During this period, Carl also wrote a masterly and massive (132 page) review on carbohydrate metabolism. He noted that most of the work had been done using whole animals, and so the underlying mechanisms were unclear.

Despite the Coris' dedication to research, it was not "all work and no play." While in Buffalo, they explored upstate New York and the shores of Lake Erie. They also took trips to New York City to see plays and had vacations in the Adirondacks, the White Mountains, and Cape Cod. They loved to read aloud, and their various tastes ran to history, biography, poetry, art and literature, and archeology. They took their adaptation to American life seriously, learning U.S. history, politics, literature, and the American outlook and way of doings things. When they visited their relatives in

Europe, they were eager to return home after witnessing the gathering clouds that presaged World War II. They became U.S. citizens in 1928.

Move to St. Louis

In 1931, Carl was surprised to receive an offer to come to Washington University School of Medicine in St. Louis as chairman of the Pharmacology Department. Philip Shaffer, the chairman of the Department of Biological Chemistry there, was a prime mover in the recruitment. Shaffer had a keen interest in the biochemistry of disease and had developed methods for purifying insulin and measuring blood glucose. Carl had had some limited teaching experience as a one-year adjunct assistant professor at the University of Buffalo. He had previously been offered a job at the University of Rochester, but he declined, in part because they required that he take speech lessons. Astonishingly, on a final visit to the University of Rochester, Gerty was taken aside and told that she was standing in the way of Carl's career and that it was un-American for a man to work with his wife! This reduced her to tears and required much assurance from Carl that their collaboration, although unusual and occasionally leading to friction, would work if there was much give and take on both sides. Both Cornell University and the University of Toronto also wanted to hire Carl, but refused to take Gerty.

When Carl arrived in St. Louis for the interview, he didn't take it particularly seriously. The evening was very hot and humid, and rather than preparing his lecture, he spent the evening under a fan reading one of Jane Austen's novels, much to the chagrin of Gerty. Prior to the visit, he was told that the chairman of anatomy was opposed to his appointment, but this gentleman was disarmed when Carl recognized that an object on his desk was the inner ear of a whale. In a shocking indictment of academic affairs at that time, Gerty could only be employed in the department as a research assistant, a nonfaculty position with a token salary, reputedly one-tenth of Carl's salary. Although Carl took the job, he pondered what exactly pharmacology was. He noted that the great chemist Justus von Liebig had stated in 1842: "Chemistry is now so intimately amalgamated with physics that it would be difficult to draw a sharp demarcation line between the two. The same intimate relationship exists between chemistry and physiology, and in half a century it will be equally impossible to separate them." The perceptive words of this great scientist assured Carl that the discipline of pharmacology would become difficult to separate from physiology and biochemistry (Figures 2&3).

Before leaving Buffalo, Carl realized from his massive review of the literature that a better way to study the processes involved in metabolism and its regulation would be to utilize isolated tissues rather than whole animals. In vivo studies were complicated by difficulties of controlling the experiments and the complex interactions between organ systems. Many of these interactions involved feedback loops that tended to counteract the primary changes. So they changed to isolated muscle preparations and disrupted tissues, and ultimately studied the enzyme catalysts that carried out the metabolic transformations. These approaches moved their research from physiology to biochemistry (Figures 2&3).

The move to Washington University was not without difficulties. The department was well equipped for neurophysiology, because the previous chair was Herbert Gasser who, with Joseph Erlanger, applied the cathode ray oscilloscope to studying nerve conduction (Figure 4). For this they won the Nobel Prize in 1944. Thus the department was well supplied, but not with the equipment needed by the Coris for their metabolic studies. This equipment had to be acquired slowly, with the money coming almost entirely from private agencies such as the Rockefeller Foundation. Carl was also required to give thirty-five to forty lectures to first-year medical students and had many laboratory classes. He did not particularly like formal teaching, and his German-inflected manner of speaking often made it difficult for the students to understand him. His focus was on research, and he was not concerned with the needs of the medical students.

One of the Coris' first discoveries in St. Louis was that, during the breakdown of muscle glycogen to lactic acid induced by epinephrine or electrical stimulation, inorganic phosphate disappeared from the muscle. This could be accounted for by an increase in hexose monophosphate (Embden ester), but not by changes in the other phosphate-containing compounds. When the Coris attended the International Congress of Physiology in Leningrad in 1935, they found that a Polish investigator, Jacob Parnas, had made similar observations. Parnas had just been appointed to the chair of biochemistry at the famous University of Lvov, which was founded in 1661. Both Parnas and the Coris decided to continue working on the problem, but using different experimental systems. The Coris used minced muscle extracted with water, but they could not get the reaction (hexose monophosphate formation) to go and almost gave up work with the system. However, the frustrating experience ended when they added a boiled muscle extract to the system. The critical factor in the extract was found to be an adenine nucleotide (AMP), which was subsequently found to stimulate phosphorylase, the enzyme responsible for the breakdown of glycogen (Figures 5&6).

In Buffalo, the Coris had developed a method for measuring both hexose (glucose) and phosphate simultaneously, and this enabled them to make quantitative measurements of hexose monophosphate produced by the breakdown of glycogen in contracting or epinephrine-stimulated muscle. Their earlier experiments demonstrated that the hexose monophosphate was formed by esterification with inorganic phosphate, and their measurements showed that it accounted for the difference between glycogen loss and lactic acid formation, leading them to propose that this phosphate ester was the source of the lactic acid. At that time (1937), the only known hexose monophosphates involved in glycolysis were hexose-6-phosphates, but when the hexose phosphate produced in muscle was analyzed carefully by the

Coris, they found a discrepancy: it was unable to reduce ferricyanide, unlike the hexose-6-phosphates. Working with a brilliant graduate student, Sidney Colowick, whose career will be discussed in chapter 2, the Coris minced 400 g of frog muscle, extracted it with water, and incubated the insoluble, glycogen-containing extract with buffer containing AMP, which accelerated the reaction. The resulting phosphate ester was crystallized as the brucine salt and subjected to elementary analysis, which was consistent with it being a hexose phosphate. However, a variety of chemical tests showed it was not glucose-6-phosphate, and the Coris deduced it to be glucose-1-phosphate. The definitive proof only came when glucose-1-phosphate was synthesized chemically and shown to be identical to the compound produced by the breakdown of muscle glycogen.

The discovery of this compound, which became known as the Cori ester, illustrates the Coris' approach to research—following a logical sequence of questions, utilizing reliable and specific analytical methods, providing complete experimental data (usually quantitative), and using chemistry to unequivocally characterize biological molecules. There was also an element of serendipity in the discovery of glucose-1-phosphate, because they extracted their muscle preparation with water to remove any preformed phosphates, but not glycogen. This procedure also extracted Mg^{2+} ions and the Mg^{2+}-requiring enzyme phosphoglucomutase This enzyme would have converted the glucose-1-phosphate to glucose-6-phosphate, thus obscuring the identification of the true product of phosphorylase action.

References

Cannon, W. B. 1929. "Organization for physiological homeostasis." *Physiol. Rev.* 9: 399–431.
Cohn, M. 1992. "Carl Ferdinand Cori." *Biogr. Mem. Natl. Acad. Sci.* 61: 78–109.
Cori, C. F. 1925. "The influence of insulin and epinephrine on the lactic acid content of blood and tissues." *J. Biol. Chem.* 63: 253–68.
———. 1925. "The fate of sugar in the animal body. 1. The rate of absorption of hexoses and pentoses from the intestinal tract." *J. Biol. Chem.* 66: 691–715.
———. 1931. "Mammalian carbohydrate metabolism." *Physiol. Rev.* 11: 143–275.
———. 1969. "The call of science." *Annu. Rev. Biochem.* 38: 1–21.
Cori, C. F., S. P. Colowick, and G. T. Cori. 1937. "The isolation and synthesis of glucose-1-phosphate." *J. Biol. Chem.* 121: 465–77.
Cori, C. F., and G. T. Cori. 1925. "The carbohydrate metabolism of tumors. I. The free sugar, lactic acid, and glycogen content of malignant tumors." *J. Biol. Chem.* 64: 11–22.
———. 1925. "The carbohydrate metabolism of tumors. II. Changes in the sugar, lactic acid, and CO_2-combining powers of blood passing through a tumor. *J. Biol. Chem.* 65: 397–405.
———. 1925. "Comparative study of the sugar concentration in arterial and venous blood during insulin action." *Am. J. Physiol.* 71: 688–707.
———. 1928. "The mechanism of epinephrine action. I. The influence of epinephrine on the carbohydrate metabolism of fasting rats, with note on new formation of carbohydrates." *J. Biol. Chem.* 79: 309–19.
———. 1928. "The mechanism of epinephrine action. II. The nfluence of epinephrine and insulin on the carbohydrate metabolism of rats in the postabsorptive state." *J. Biol. Chem.* 79: 321–41.

———. 1928. "The mechanism of epinephrine action. III. The influence on the utilization of absorbed glucose." *J. Biol. Chem.* 79: 343–55.

———. 1929. "The mechanism of epinephrine. IV. The influence of epinephrine on lactic acid production and blood sugar utilization." *J. Biol. Chem.* 84: 683–98.

———. 1929. "Glycogen formation in the liver from d- and l-lactic acid." *J. Biol. Chem.* 81: 389–403.

———. 1931. "The influence of epinephrine and insulin injections on hexosemonophosphate content of muscle." *J. Biol. Chem.* 94: 581–91.

———. 1936. "Mechanism of formation of hexosemonophosphate in muscle and isolation of a new phosphate ester." *Proc. Soc. Exp. Biol. Med.* 34: 702–05.

Cori, G. T., and C. F. Cori. 1931. "A method for the determination of hexosemonophosphate in muscle." *J. Biol. Chem.* 94: 561–79.

———. 1933. Changes in hexose phosphate, glycogen and lactic acid during contraction and recovery of mammalian muscle." *J. Biol. Chem.* 99: 493–505.

Cori, G. T., C. F. Cori, and K. W. Buchwald. 1930. "The mechanism of action of epinephrine. V. Changes in liver glycogen and blood lactic acid after injection of epinephrine and insulin." *J. Biol. Chem.* 86: 375–88.

Feldman, B. 2000. *The Nobel Prize.* New York: Arcade Publishing, Inc.

Hegnauer, A. H., and G. T. Cori. 1934. "The influence of epinephrine on chemical changes in isolated frog muscle." *J. Biol. Chem.* 105: 691–703.

Kalckar, H. M. 1983. "The isolation of Cori ester "the Saint Louis Gateway" to a first approach of a dynamic formulation of macromolecular biosynthesis" *Comprehensive Biochemistry.* Vol. 35. Edited by G. Semenza, 1–24. Amsterdam: Elsevier Science.

Krebs, H. A. 1972. "Otto Heinrich Warburg." *Biogr. Mems Fell. R. Soc.* 18: 628–99.

Larner, J. 1992. "Gerty Theresa Cori." *Biogr. Mem. Natl. Acad. Sci.* 61: 110–35.

Macleod, J. J. R. 1929. "Physiology of glycogen." *Lancet ii* 1(55): 107.

Randle, P. J. 1986. "Carl Ferdinand Cori." *Biogr. Mems. Fell. R. Soc.* 32: 66–95.

2

Sidney Colowick

Their First Graduate Student

Research in the Cori Laboratory

Sidney Colowick was born on January 12, 1916, and raised in St. Louis, Missouri. His parents were Jews who had emigrated from Lithuania and ran a dry cleaning business. Colowick graduated from Washington University in 1936 at age twenty with a degree in chemical engineering. Due to the continuing Great Depression, jobs were hard to find, and he took a civil service test, which qualified him as a shrimp inspector, although what that would entail mystifies me! His parents discouraged him from taking such a position in New Orleans because of his youth and the reputation of that city. He learned of an opening in the Cori laboratory, and Carl accepted him on a four-week trial basis, although at much lesser pay that he would have made as a shrimp inspector and despite the fact that he dropped a desiccator containing valuable reagents just outside Carl Cori's office. The Coris soon recognized Colowick's talent and made the position permanent. Gerty looked upon him as a son, although the feeling was not reciprocated. Colowick would be associated with the Coris for ten years, publishing his first paper (on glucose-1-phosphate) at age twenty-one. Nathan Kaplan, with whom Colowick would become closely associated, tells how Colowick described himself as the "meat" in the "Cori sandwich" and further notes that the shrimp inspector who knew nothing about biochemistry established himself in a few years as one of the most promising young people in the field.

Colowick was not only a brilliant and creative scientist, but also a compassionate and warm person. These attributes would remain with him throughout his career in which he was always open to provide guidance and assistance to other scientists. Despite his relative youth and lack of formal training in biochemistry, he was a major contributor to the Coris' research program. He was at first the only graduate student in the laboratory, but was exposed to many senior scientists, almost all of whom would move on to distinguished careers and some of whom would become Nobel Laureates. Thus he was the beneficiary of a laboratory filled with brilliant scientists. Because of his intellectual gifts and experimental skills, he collaborated with many of these scientists as well as with the Coris. One of them was Herman Kalckar, whose career is described in chapter 3. Kalckar stated that Colowick, "although only 25 years old, had achieved wisdom as well as a warm sense of humor and brilliant analytical insight." Kalckar and Colowick worked together on the enzyme hexokinase, which converts glucose to glucose-6-phosphate in the first step of the glycolytic pathway. They were troubled by the fact that when crude hexokinase preparations were compared with pure preparations, ATP was converted to AMP instead of ADP. This led to the discovery of a contaminating enzyme that was later shown to convert two molecules of ADP to ATP and AMP. The enzyme was later called adenylate kinase and plays an important role in "rescuing" AMP in cells by converting it to ADP, which is then converted to ATP by various mechanisms.

In addition to his discovery that glucose-1-phosphate was the product of glycogen breakdown by phosphorylase, which was described in the previous chapter, Colowick was crucially involved in the identification of phosphoglucomutase (which converts glucose-1-phosphate to glucose-6-phosphate) and in the recognition that AMP is an activator of phosphorylase. These discoveries formed the basis for his PhD, which was awarded in 1942. As described in a later chapter, working in the Cori laboratory, Victor Najjar isolated, purified, and crystallized phosphoglucomutase and defined its properties, which was no mean feat at that time. Colowick was also the sole author of a paper dealing with the synthesis of two other sugar phosphates, namely mannose-1-phosphate and galactose-1-phosphate. An important collaboration was with Earl Sutherland, who spent a period (1940–1942) in the Cori lab while a medical student at Washington University and later as a research fellow. Sutherland's career is described in chapter 7. He and Colowick became close friends and worked together on several problems. They published a classic paper on the synthesis of glycogen from glucose. The importance of this was that they used purified enzymes and that it was the first in vitro synthesis of a macromolecule. Colowick's later focus at St. Louis was on glucose phosphorylation in tissue extracts, which led to studies of hexokinase that he obtained pure from yeast. This interest would expand later in his career. He studied the effects of insulin and extracts of the adrenal and anterior pituitary glands on the activity of hexokinase in vitro. The reported stimulatory effects of insulin on hexokinase created great excitement in the field, but sadly could not be confirmed, and it appeared that one of his collaborators had fabricated the data. This was a severe psychological blow to Colowick and the Coris.

Subsequent History

In 1946, Colowick moved to the Public Health Institute of the City of New York, replacing Herman Kalckar, who returned to Denmark. This was an unhappy period when he spent time trying to repeat the questionable results with insulin and hexokinase to no avail. Despite his failure, he showed great equanimity and exhibited no animosity to his former colleague. He moved to the Department of Biochemistry at the University of Illinois in Chicago, but this was again an unhappy experience. This was because he was saddled with the task of preparing a new laboratory syllabus for more than three hundred students. He did this all by himself, just keeping one experiment ahead of the students. The effort was so stressful that he developed a duodenal ulcer and had to be hospitalized. One positive aspect was that during his stay, he met up with Nathan (Nate) Kaplan, who had joined the faculty. Kaplan had obtained his PhD from the University of California, Berkeley, after spending some time as a chemist on the Manhattan Project. He was a postdoctoral fellow under Fritz Lipmann at the Massachusetts General Hospital (MGH), where they isolated and studied coenzyme A, which plays a crucial role in many aspects of metabolism. Lipmann won the Nobel Prize in 1953 for this discovery. Colowick and Kaplan commenced a fruitful collaboration when they both moved to the new McCollum-Pratt Institute of Johns Hopkins University under the direction of William McElroy, who discovered the luciferin-luciferase system which generates the light of fireflies. The system utilizes ATP and has formed the basis for a very sensitive method for measuring this nucleotide. McElroy created an environment more favorable to research, and Colowick and Kaplan began studies of pyridine nucleotides. These nucleotides, which are now called NADH and NADPH, are cofactors for many dehydrogenases that play crucial roles in oxidative processes in the cell. The two researchers studied the chemistry of the nucleotides and the enzymes that utilized them. Importantly, Colowick, working with Maynard Pullman and Anthony San Pietro using deuterium as a tracer, determined the correct structure of NADH and the site at which the oxidation/reduction occurs. The redoubtable Otto Warburg, who had discovered this coenzyme, had gotten the structure wrong, and Birgit Vennesland had identified the wrong position for the oxido-reduction. Pullman would later make contributions in the area of oxidative phosphorylation, and San Pietro would discover ferredoxin a protein involved in the light reactions of photosynthesis.

The proximity of Baltimore to the biomedical research complex at the National Institutes of Health (NIH) in Bethesda provided the researchers at the McCollum-Pratt Institute with the opportunity to interact with other young biochemists, in particular, Arthur Kornberg, Bernard Horecker, Herbert Tabor, Alton Meister, Leon Heppel, Christian Anfinsen, and Earl and Thressa Stadtman. All of them would achieve leadership positions in American biochemistry, and two would become Nobel Laureates. In one of their major achievements, Colowick and Kaplan were approached by Academic Press to edit a series of volumes dedicated to biochemical methodology. The series, begun in 1955, was called *Methods in Enzymology* and rapidly became known as the biochemists' bible. Colowick coedited the series until his death in 1985, and it continues today reaching volume 500 at the time of this writing. Colowick left Baltimore in 1959 to take a position as Charles Hayden-American Cancer Society Professor in the Department of Microbiology at Vanderbilt University School of Medicine (Figure 7), which was chaired by Victor

Najjar who had also spent time in the Cori department. At Vanderbilt, Colowick continued his work on hexokinase and also focused on hexose transport in cultured animal cells. His research displayed his characteristic creativity, but his contributions went beyond basic science to becoming a force for the betterment of Vanderbilt. Thus he helped attract Earl Sutherland there and gave freely of his time to revise the papers of students and colleagues. He could be counted on to ask incisive questions at seminars or formal talks, whether they were given by graduate students or Nobel Laureates. His editorial skills were legendary, and it is said that most colleagues at Johns Hopkins and Vanderbilt had him review papers prior to submission. He received numerous honors, including election to the American Academy of Arts and Sciences and the National Academy of Sciences. He died on January 9, 1985, just two months after his mentor, Carl Cori.

At his memorial service, Alexander Heard, Chancellor Emeritus of Vanderbilt University, who was noted for his distinguished mien and eloquence, said: "The heart of the university in western civilization is its duty to inquire and discover, and to interpret and communicate a useful harvest. In these central missions of the most influential institution of the twentieth century, the university, Sidney Paul Colowick excelled. He was a person of science, of the intellect, of the university, of the eternal human search to know and understand."

References

Berger, L., M. W. Slein, S. P. Colowick, and C. F. Cori. 1946. "Isolation of Hexokinase from Baker's Yeast. *J. Gen. Physiol.* 29: 141–42.

Colowick, S. P. 1938. "Synthetic Mannose-1-Phosphoric Acid and Galactose-1-Phosphoric Acid." *J. Biol. Chem.* 124: 557–58.

Colowick, S. P., and E. W. Sutherland. 1942. "Polysaccharide Synthesis from Glucose by Means of Purified Enzymes." *J. Biol. Chem.* 144: 423–37.

Colowick, S. P., M. S. Welch, and C. F. Cori. 1940. "Phosphorylation of Glucose in Kidney Extract." *J. Biol. Chem.* 133: 359–73.

Colowick, S. P., H. M. Kalckar, and C. F. Cori. 1941. "Glucose Phosphorylation and Oxidation in Cell-Free Tissue Extracts." *J. Biol. Chem.* 137: 343–56.

Cori, C. F., S. P. Colowick, and G. T. Cori. 1937. "The Isolation and Synthesis of Glucose-1-Phosphoric Acid." *J. Biol. Chem.* 121: 465–77.

Cori, G. T., S. P. Colowick, and C. F. Cori. 1938. "The Formation of Glucose-1-Phosphoric Acid in Extracts of Mammalian Tissues and of Yeast." *J. Biol. Chem.* 123: 375–80.

———. 1938. "The Action of Nucleotides in the Disruptive Phosphorylation of Glycogen." *J. Biol. Chem.* 123: 381–89.

———. 1938. "The Enzymatic Conversion of Glucose-1-Phosphate Ester to 6-Ester in Tissue Extracts." *J. Biol. Chem.* 124: 543–55.

Inman, W. H., and S. P. Colowick. 1985. "Stimulation of Glucose Uptake by Transforming Growth Factor Beta: Evidence for the Requirement of Epidermal Growth Factor-Receptor Activation." *Proc. Natl. Acad. Sci. U.S.A.* 82: 1346–49.

Kalckar, H. M. 1991. "50 Years of Biological Research—From Oxidative Phosphorylation to Energy Requiring Transport Regulation." *Ann. Rev. Biochem.* 60: 1–38.

Kaplan, N. O. 1981. "Sidney P. Colowick." *Methods Enzymol.* 113: xvii–xxii.

Kaplan, N. O., S. P. Colowick, and E. F. Neufeld. 1952. "Pyridine Nucleotide Transhydrogenase: II. Direct Evidence for the Reaction and Mechanism of the Transhydrogenase." *J. Biol. Chem.* 195: 107–19.

Najjar, V. A. 1948. "The Isolation and Properties of Phosphoglucomutase." *J. Biol. Chem.* 175: 281–90.

Price, W. H., C. F. Cori, and S. P. Colowick. 1945. "Effect of Anterior Pituitary Extract and of Insulin on the Hexokinase Reaction." *J. Biol. Chem.* 160: 633–34.

Pullman, M. E., A. San Pietro, and S. P. Colowick. 1954. "On the Structure of Reduced Diphosphopyridine Nucleotide." *J. Biol. Chem.* 206: 129–41.

Schulze, I. T., and S. P. Colowick. 1969. "The Modification of Yeast Hexokinases by Proteases and Its Relationship to the Dissociation of Hexokinase into Subunits." *J. Biol. Chem.* 244: 2306–16.

Trayser, K. A., and S. P. Colowick. 1961. "Properties of Crystalline Hexokinase from Yeast. IV. Multiple Forms of the Enzyme." *Arch. Biochem. Biophys.* 94: 177–81.

3

Herman Kalckar

The Great Dane

Early Career

Herman Moritz Kalckar spent three years (1940–1943) in the Cori laboratory, working principally with Sidney Colowick. Kalckar was born in Copenhagen into a middle-class Jewish family on March 26, 1908. His mother read widely in French and German (e.g., Flaubert, Proust, Goethe, Heine) and spoke both languages. His father, a business consultant and broker, was not wealthy and refused to profit from World War I. He read Jewish and Yiddish humorists and wrote enthusiastically about attending the world premiere of Henrik Ibsen's *A Doll's House* at the Royal Theatre in 1879. Herman's brother Fritz was a gifted physicist and protégé of Niels Bohr, who won the Nobel Prize in 1922 for his work on atomic structure. Fritz's early death from status epilepticus was a great blow to the family. Herman attended a local school and described himself as an erratic pupil who occasionally felt miserable. The headmaster was a world-renowned scholar of Greek and impressed Herman greatly. He described his physics instructor as a formidable and passionately devoted teacher who wrote a very concentrated textbook that emphasized mathematics, especially calculus. The importance of these gifted high school teachers in the careers of many of the scientists described in this book is a recurring theme. Kalckar was impressed that the renowned physiologist August Krogh came to his high school to demonstrate the principles of human physiology. Krogh won the Nobel Prize in 1920 for his work on the regulation of blood flow through capillaries. In an interesting example of the interrelations between scientists, Krogh's mentor had been Christian Bohr, the father of Niels Bohr, with whom Kalckar's brother Fritz had worked. Christian Bohr had discovered the Bohr effect in which CO_2 promotes the dissociation of O_2 from hemoglobin, thus facilitating the release of O_2 to active tissues.

Kalckar studied medicine at the University of Copenhagen and, after graduation in 1933, began work on a PhD under the physiologist Ejnar Lundsgaard. Lundsgaard had shown that frog muscles could contract in the absence of glycolysis (blocked by iodoacetate), and this was a shock to Otto Meyerhof's group in Berlin-Dahlem, which included Fritz Lipmann, who believed that the conversion of glycogen to lactic acid was the link between metabolic energy generation and muscle contraction. Meyerhof had received the Nobel Prize in 1922 with the English physiologist Archibald Vivian Hill (usually known as A. V. Hill) for his work in this area. The dilemma was solved when two scientists at Harvard, Cyrus Fiske and Yellagaprada SubbaRow, developed a method for measuring phosphate, leading to the discovery of creatine phosphate in muscle. This compound had been identified earlier by Grace and Philip Eggleton at the University of Cambridge, but its chemical nature had not been defined. Karl Lohmann, working in Meyerhof's group, had discovered ATP. This discovery was of fundamental importance since ATP plays a central role in transferring energy to hundreds of cellular processes. Lundsgaard's results were explained when it was realized that it was the creatine phosphate store which, by conversion to ATP via ADP, provided the energy required for muscle contraction in the absence of glycolysis. There was a lively competition between Lohmann and Fiske and SubbaRow as to who discovered these phosphate esters first.

Influence of German Jewish Scientists

Fritz Lipmann came to Copenhagen from the Meyerhof laboratory in 1932 because his Jewish heritage prevented him from being able to work in Germany. This was the start of a lifetime relationship between Lipmann and Kalckar. Lipmann was born in Königsberg in East Prussia and studied medicine in Berlin and, after a brief stay in Amsterdam, he joined Meyerhof's group and then moved to the Carlsberg Laboratory in Copenhagen.

Lipmann encouraged Kalckar to study the work of Hans Adolf Krebs, the discoverer of the citric acid cycle. Also known as the Krebs cycle, it is a cyclic series of enzymatic reactions that generate most of the ATP in cells under aerobic conditions (Krebs would win the Nobel Prize in Physiology or Medicine with Lipmann in 1953). Like Otto Warburg, Hans Krebs was a towering figure in the field of metabolism. He was born in Hildesheim, Germany, in 1900 and studied medicine at the Universities of Göttingen and Freiburg before obtaining a PhD at the University of Hamburg. He then worked with Otto Warburg until 1930, when he returned to clinical work at the University of Freiburg and discovered the urea cycle (a series of reactions by which the amino groups of amino acids are converted to urea). Because of his Jewish heritage, his appointment was terminated, and he emigrated to England at the invitation of Frederick Gowland Hopkins (Nobel Prize winner in 1929 for his work on vitamins), who was head of the Biochemistry Department at Cambridge University. Krebs then moved to the University of Sheffield, where he carried out classic studies that led to the formulation of the citric acid cycle. In 1954, he was appointed Whitley Professor and head of the Biochemistry Department at

Oxford University, despite some lingering anti-German sentiment. Krebs' brilliance was undeniable, but he had a rather ascetic and humorless demeanor.

Like Warburg, Otto Meyerhof had a profound effect on the researches of the Cori laboratory. He was born in Hanover, Germany in 1884 and grew up in Berlin. He graduated in Medicine from the University of Heidelberg and did research in metabolism there. He met Otto Warburg, whose ideas and approaches inspired him to focus his career on physiological chemistry. In 1912, Meyerhof took a position at the University of Kiel and provided one of the first adaptations of the laws of thermodynamics to physiological chemistry. He recognized that after the input of energy in the form of food, it is transformed through a series of steps and finally dissipated as heat. Soon after the end of World War I, Meyerhof began collaborating with A. V. Hill to decipher muscle metabolism in terms of heat production, mechanical work, and cellular chemical reactions. He determined that glycogen is converted to lactic acid in the absence of oxygen, but that in the presence of oxygen, it is mostly reconverted to glycogen. These results were later termed the Pasteur-Meyerhof effect, and their relevance to the findings of the Coris is obvious. After winning the Nobel Prize, Meyerhof went to Berlin-Dahlem and then to a newly formed Kaiser Wilhelm Institute in Heidelberg. He undertook the task of dissecting glycolysis into its component reactions and was very successful in accomplishing this, identifying more than one-third of the enzymes involved. Importantly, Meyerhof made the first associations between the uptake of phosphate during the breakdown of glucose and glycogen to lactic acid and the synthesis of ATP.

Like Krebs and Lipmann, Meyerhof had to leave Germany because of Hitler's rise to power. He took a position in Paris in 1938, and, when the Nazis invaded France, he left for the United States to assume the post of research professor of physiological chemistry, created for him by the University of Pennsylvania. He continued his research there until his death on October 6, 1951.

Fritz Lipmann, like Meyerhof, was another scientist and biochemist who emphasized the central role of ATP as an "energy-rich" phosphate ester whose breakdown drove not only muscle contraction but many other energy-requiring cellular processes. Otto Meyerhof and Karl Lohmann, working in the Kaiser Wilhelm Institute in Berlin-Dahlem, had demonstrated that ATP was produced during glycolysis, but how it was produced during oxidation of substrates remained obscure. This oxidative process is of great importance since it generates about seventeen times more ATP from glucose than anaerobic glycolysis (i.e., that occurring in the absence of O_2). Encouraged by Lipmann, Kalckar proceeded to study this process—termed oxidative phosphorylation—in kidney cortex extracts. Kalckar found that in the presence of oxygen (aeration), phosphorylation of various substrates occurred, but this did not happen under anaerobic conditions unless dicarboxylic acids such as fumarate or malate were added. Under these conditions, a novel phosphate ester was formed. When Lohmann and Meyerhof discovered phospho*enol* pyruvic acid as a component of the glycolytic pathway, Kalckar realized that this was the compound produced in his experiments. Unknown to him, the conversion of oxaloacetic acid (which is derived from fumarate and malate in the citric acid cycle and from pyruvate by pyruvate carboxylase) to phospho*enol* pyruvate would become recognized as a key reaction in the gluconeogenic pathway by which lactic acid is converted to glucose.

Kalckar's studies on oxidative phosphorylation proved to be resistant to experimentation, and there was enormous controversy and rancor among the various groups studying the process. It had been discovered in 1932 by the Russian scientists Wladimir Engelhardt and his wife Malitza Lyubimova, and there was probably no greater dispute among biochemists than that concerning its mechanism. The principal players were Efraim Racker, Britton Chance, David Green, Paul Boyer, Henry Lardy, Albert Lehninger, William Slater, and Lars Ernster. None of them were "wilting violets," and their exchanges were protracted and fiery during conferences, with none of the protagonists conceding defeat. Two decades were to pass before the English scientist Peter Mitchell developed the novel, and not readily accepted, chemiosmotic theory by which a gradient of H^+ ions across a mitochondrial membrane drives ATP production. Mitchell received the Nobel Prize for this in 1978.

First Visit to the United States

In 1939, Kalckar defended his PhD at the University of Copenhagen and decided to go to the United States. One of his mentors, Kaj Linderstrøm-Lang at the Carlsberg Laboratory, suggested he go to the California Institute of Technology in Pasadena, California. On his way there, Kalckar made a brief stop at the Cori laboratory. He found St. Louis to be a bleak and sooty city, but the Coris were greatly interested in his work. They were concerned, however, because they couldn't repeat Kalckar's experiments on oxidative phosphorylation. The problem was solved when he told them to use a simple shaking technique, which provided better oxygenation—the key experiment was conducted by Sidney Colowick. He remembers Kalckar saying "shake it!" And everything was ok.

At Pasadena, Kalckar made friends with Max Delbrück (Nobel Prize in 1969 for mechanisms of viral replication) and David and James Bonner. The Bonner brothers were both plant physiologists, with David later switching to the study of genetics of the yeast *Neurospora*. David died prematurely at the age of forty-eight of the effects of massive radiation therapy for Hodgkin's lymphoma. Kalckar describes the Bonner brothers as great scientific colleagues. One of their important contributions was to introduce him to a superb mechanic who kept his car running. This was of significance because, during a visit by the Swedish scientist Hugo Theorell (Nobel Prize in 1955 for studies of oxidative enzymes), Kalckar had to drive Theorell up the snow-covered mountains to the Mt. Wilson observatory. Theorell introduced Kalckar to Linus Pauling (Nobel Prize in chemistry in 1954), and Kalckar incorporated Pauling's ideas about resonance in a review he was writing. If interacting with these Laureates was not enough, he and his musician wife Vibecke were invited to have Christmas dinner in 1939 at the home of Thomas Hunt Morgan (Nobel Prize in 1933 for the role of chromosomes in heredity) in the coolness of Pacific Grove near Monterey, California.

In 1940, the Kalckars traveled by train to St. Louis, stopping briefly in New Orleans. St. Louis was depressing as before, but Herman was excited by the new type of biochemistry being practiced in the Cori laboratory. As described above, he joined up with Sidney Colowick in studies of hexokinase, which would lead to the discovery of adenylate kinase. In his autobiography, Kalckar describes his work with Colowick as some of his happiest months in research. When Colowick was involved in defense-related research in 1942, Kalckar continued alone on the hexokinase project. Of course, it was impossible for him to return to Denmark at this time because of the Nazi occupation, and he knew nothing of the fate of his brother and friends who had resisted the Germans.

In 1943, he received an invitation from Oliver Lowry to come to the Public Health Institute of the City of New York. The pride of the laboratory was a new Beckman UV spectrophotometer, which Kalckar described as his "Stradivarius." This piece of equipment would transform biochemical research across the world. Lowry was a superb methodologist who invented the famous Lowry method for measuring proteins. His publication of the method in the *Journal of Biological Chemistry* became the most cited paper in publishing history. Kalckar's project was to study the enzymatic synthesis of purine nucleosides, and he made the important discovery that the phosphorolytic cleavage of nucleosides was similar to that of glycogen. The reaction was found to be reversible and to synthesize nucleosides. Later work showed that the reaction was more concerned with degradation rather than synthesis.

Return to Denmark

The Kalckars returned to Copenhagen in 1946 where Herman continued his work on nucleosides and nucleotides in a new laboratory arranged for him by his former graduate school mentor, Ejnar Lundsgaard. Lundsgaard attracted many brilliant predoctoral and postdoctoral students, including Hans Klenow, who originated the Klenow polymerase method for making mutations, and Morris Friedkin from Albert Lehninger's laboratory in Chicago, who was instrumental in the studies of the phosphorolysis of nucleosides. Klenow and another student studied the enzymes xanthine oxidase and xanthopterin oxidase, and this necessitated their isolation from milk using a large Danish milk centrifuge. Kalckar later decided that the project did not deserve to be pursued! On the other hand, he recommenced work on the nucleotide phosphorylase project in collaboration with Friedkin. In the course of this work, he encountered James Watson, who would later win the Nobel Prize in 1962 for elucidating the structure of DNA, together with Francis Crick and Maurice Wilkins. Kalckar described Watson as a shy young man—a characterization that did not persist in his career. Another person with whom Kalckar interacted was Alexander Todd, who was chair of chemistry at Cambridge. Todd was a tall Scot with an imperious presence who had married the daughter of the Nobel Laureate Sir Henry Dale. Todd won the Nobel Prize in 1957 for his work on the structure of nucleotides and nucleosides and would be knighted and later made Baron Todd of Trumpington. He was elected to the Royal Society and made president. I remember being a graduate student in New Zealand and hearing him state with assurance that DNA was so complex that its structure would never be solved! Two years later, he was proven wrong.

Paul Berg, who won the Nobel Prize in Chemistry for his work on recombinant DNA, visited the Kalckar laboratory as a postdoctoral fellow in 1952. His PhD mentor at Western Reserve University, Harland Wood, had arranged for him to go to work with the Coris. Much to Wood's annoyance, Berg refused to go to St. Louis because of the vestiges of racial segregation and the notoriously torrid summers. Wood considered the idea of turning down a career-making opportunity because of an aversion to a city to be shortsighted, if not foolish. Berg describes the Kalckar group as being an interesting collection, He describes Watson's visit as being short because of his disdain for biochemistry. and he describes Kalckar as having a captivatingly charming, buoyant, and fun-loving manner, but being notoriously difficult to understand. He learned later that it wasn't just his Danish-like English that confused people, because even the Danes found it difficult to follow Kalckar when he spoke his native language. Berg later went to St. Louis to work with Arthur Kornberg. Kornberg's discovery of the enzyme (DNA polymerase) that synthesizes DNA is described later.

In his autobiography, Kalckar recounts that Linderstrøm-Lang invited the great Otto Warburg to Copenhagen to give two lectures at the Carlsberg Laboratory. Warburg expressed an interest in seeing Elsinore and Hamlet's castle, and Kalckar took him on the excursion, which turned out to be very pleasant. However, he was surprised when Warburg asked him: "Why do I encounter so much alienation from British and American biochemists?" Kalckar answered this difficult question by quoting one of his admirers who treasured Warburg's monumental prose, but occasionally worried about some of his opinionated footnotes. "Be honest with me," Warburg said, "give me one example." "Ok." said Kalckar. "Why did you call the late Sir Frederick Gowland Hopkins a romantic?" "Did I really?" responded Warburg. "Yes, in one of your footnotes," replied Kalckar. Warburg assured him that he had always been an admirer of Hopkins. Kalckar considered this open confession from one of the giants of science to be remarkable, and he communicated it to the Coris who were old friends of Warburg.

Return to the United States

In the 1950s, Kalckar turned his attention to the enzymology of galactose metabolism. In this, he was influenced by the pioneering studies of Luis Leloir, who also spent time with the Coris and whose work will be described later. In the course of his work, Kalckar found an alternative route for the synthesis of galactose in yeast. This involved the interconversion of UDP glucose and UDP galactose catalyzed by uridyl transferase. In 1952, he returned to the United States as part of the Visiting Scientists Program at NIH and later achieved a permanent appointment. He worked with Jack Strominger (who elucidated the bactericidal action of penicillin) and Julius Axelrod (who won the Nobel Prize for his work on catecholamine neurotransmitters) to study a novel pathway involving UDP-glucose by which glucose could be converted to glucuronic acid. In addition, in conjunction with the noted gastroenterologist Kurt Isselbacher, Kalckar developed an assay for uridyl transferase, which was used to show that it was deficient in the most serious form of galactosemia, a disease in which galactose-1-phosphate accumulates in cells throughout the body and can cause severe gastrointestinal symptoms. The test is now used to screen newborn infants.

Just as Sidney Colowick and Nathan Kaplan had been invited by William McElroy to take positions at the McCollum-Pratt Institute at Johns Hopkins, so also was Kalckar recruited there in 1958. In this year, he published a novel proposal to test the fallout from nuclear tests by measuring the ^{90}Sr content in the milk teeth of children. This test revealed high levels of the isotope in children born in 1950, which declined when atmospheric testing was stopped. At Hopkins, Kalckar studied galactose metabolism in mutant lines of the bacterium *Escherichia coli* (*E. coli*). In 1960, he moved to the Huntington Laboratories at the Massachusetts General Hospital (MGH) in Boston. There, Kalckar succeeded Fritz Lipmann as head of the Biochemical Research Laboratory. He was recruited there by Paul Zamecnik, head of the laboratories, and by Eugene Kennedy, newly appointed chair of biochemistry at Harvard Medical School. Both of these scientists had made classic contributions to the understanding of protein synthesis and fat metabolism, respectively. While there, Kalckar received some visitors from the Carlsberg Laboratory who were accompanied by Danish lady whom he had met and admired thirty years previously and whose husband had died in 1960—Kalckar and she married in 1968. His work at MGH continued to be focused on *E. coli* mutants with defects in galactose metabolism and was later extended to yeast and then to a form of chemotaxis in which bacteria are attracted to galactose. This was shown to be due to the presence of receptors on the surface of the cells that bound galactose with high affinity. In 1973, he was attracted to the field of tumor biology, specifically the enhanced transport of hexoses into transformed (tumor-like) cells and its regulation.

Kalckar gained a worldwide reputation and, in his autobiography, he described in particular his visits to the Soviet Union, Japan, and China where every stop was filled with intriguing episodes. Despite the fact that he was invited to present numerous seminars and lectures, he was not a particularly effective speaker. He tended to mumble, as noted by Paul Berg and complained about by Max Delbrück, and there were often gaps in his presentations, which the audience was expected to fill.

Eugene Kennedy describes him as beginning a new topic *in media res* (in the middle of the thing, i.e., without a preamble). He retained his Danish accent, to which part, but not all, of his problem of communication was attributed.

Kalckar divorced his first wife, musician Vibeke Meyer, in 1950 and married developmental biologist Barbara Wright, with whom he had three children to whom he was very devoted. In 1968, he married Agnete Laursen, an interior designer. Kennedy described their home as a focus of warmth and hospitality for friends and colleagues. In the final phase of his career, he was a visiting professor in the Chemistry Department of Boston University, where he continued working on glucose transport. He died in Cambridge, Massachusetts, on May 17, 1991. Because of his lengthy list of discoveries, some people wonder why he didn't receive the Nobel Prize. I think it was because of his personality and intense curiosity about a wide range of topics that prevented him from focusing on one or two questions. He was, however, elected to the U.S. National Academy of Sciences, the Royal Danish Academy, and the American Academy of Arts and Sciences, and received several honorary degrees from prestigious universities.

References

Axelrod, J., H. M. Kalckar, E. S. Maxwell, and J. L. Strominger. 1957. "Enzymatic Formation of Uridine Diphosphoglucuronic Acid." *J. Biol. Chem.* 224: 79–90.
Berg, P. 2003. "Moments of Discovery: My Favorite Experiments." *J. Biol. Chem.* 278: 40417–24.
Engelhardt, W. A., and M. N. Ljubimova. 1939. "Myosin and Adenosine Triphosphatase." *Nature* 144: 668–72.
Fiske, C. H., and Y. SubbaRow. 1929. "Phosphocreatine." *J. Biol. Chem.* 81: 629–79.
Isselbacher, K. J., E. P. Anderson, K. Kurahashi, and H. M. Kalckar. 1956. "Congenital Galactosemia, a Single Enzymatic Block in Galactose Metabolism." *Science* 123: 635–36.
Kalckar, H. M. 1938. "Formation of a New Phosphate Ester in Kidney Extracts." *Nature* 142: 871.
———. 1945. "Enzymatic Synthesis of a Nucleoside." *J. Biol. Chem.* 158: 723–24.
———. 1958. "An International Milk Teeth Radiation Census." *Nature* 182: 283–84.
———. 1971. "The Periplasmic Galactose Binding Protein of Escherichia coli." *Science* 174: 557–65.
———. 1991. "50 Years of Biological Research from Oxidative Phosphorylation to Energy-Requiring Transport Regulation." *Ann. Rev. Biochem.* 60: 1–37.
Kalckar, H. M., B. Braganca, and A. Munch-Petersen. 1952. "Uridyl Transferases and Formation of UDP-galactose." *Nature* 172: 2–4.
Kennedy, E. P. 1996. "Herman Moritz Kalckar." *Biogr. Mem. Natl. Acad. Sci. U.S.A.* 69: 149–64.
Lipmann, F. 1984. "A Long Life in Times of Great Upheaval." *Ann. Rev. Biochem.* 53: 1–34.
Lohmann, K. 1929. "Über die Pyrophosphatfraktion im Muskel." *Naturwissenschaften* 17: 624–25.
Meyerhof, O., and R. Junowicz-Kocholaty. 1943. "The Equilibria of Isomerase and Aldolase, and the Problem of the Phosphorylation of Glyceraldehyde Phosphate." *J. Biol. Chem.* 149: 72–92.
Nachmansohn, D., S. Ochoa, and F. A. Lipmann. 1952. "Otto Meyerhof." *Science* 115: 365–69.
Quayle, J. R. 1982. "Obituary. Sir Hans Krebs. 1900–1981." *J. Gen. Microbiol.* 128: 2215–20.

4

Severo Ochoa

Spanish Genius

Early Education

Severo Ochoa stands out among the cavalcade of brilliant scientists who worked in the Cori laboratory (Figure 9). He was born in 1905 in Luarca, a small town in the province of Asturias on the northern coast of Spain. This is on a most attractive site at the mouth of the Río Negro with a fishing harbor and a lighthouse. Ochoa was the youngest of seven children, and his father was a lawyer and a businessman. His interest in the natural sciences was sparked by a bright young teacher of chemistry. Ochoa entered the Medical School of the University of Madrid in 1923, where he hoped to study under the famous neuroanatomist Santiago Ramón y Cajal, who had won the Nobel Prize in 1906 for his work on the structure of the nervous system, but Ochoa was greatly disappointed to hear that Ramón y Cajal had just retired. However, Ochoa was particularly influenced by a young professor of physiology, Juan Negrín, who encouraged him to read Ramón y Cajal's papers and autobiography. Negrín offered Ochoa the opportunity to do research at an institution that Ochoa describes as being like an Oxford College. He lived there in a student residence (La Residencia) that provided cultural activities, including music recitals and lectures on art, literature, history, and philosophy. In later years, the poet and playwright Frederico García Lorca, the painter Salvador Dalí, and the film director Luis Buñuel lived there.

Ochoa's project involved studies of creatine and creatinine in muscle, and he and a fellow classmate devised a micromethod for determining tissue levels of creatine. They sent the method to the *Journal of Biological Chemistry*, little believing it would be considered for publication. It was published in 1929, requiring minimal revision and with the English untouched! Although they found that stimulated muscle released creatine, whereas resting muscle did not, they failed to measure phosphate release and missed the connection to creatine phosphate.

Research in Germany and Britain

When he neared the end of his medical training, Ochoa decided to go to the Kaiser Wilhelm Institute at Berlin-Dahlem, which he recognized was the world center of research on carbohydrate metabolism and energy. As noted in previous chapters, in various divisions of the Institute were Otto Warburg, Otto Meyerhof, and a host of very able assistants and students (Karl Lohmann, Fritz Lipmann, and David Nachmansohn). Initially, Ochoa's German was limited, but Meyerhof spoke English and Ochoa eventually mastered the German language. In 1929, Meyerhof moved to Heidelberg, where the Kaiser Wilhelm Gesellschaft (Society) had built a beautiful new building. Ochoa joined him there after taking final examinations for his MD. As described in the previous chapter on Herman Kalckar, Meyerhof believed that the energy used in muscle contraction came from the conversion of glycogen to lactic acid, but the findings of Ejner Lunsgaard that muscles could contract in the absence of lactic acid formation did not support this. Neither was it supported by the German physiological chemist Gustav Embden's measurements of lactic acid in contracting and relaxing muscles. Embden worked at the Physiological Institute at the University of Frankfurt and, together with Meyerhof, had made major contributions to the identification of the steps involved in the conversion of glycogen to lactic acid. This led this series of enzymatic steps to be called the Embden-Meyerhof or glycolytic pathway. Even though it was later realized that the energy source in Lunsgaard's experiments was creatine phosphate, the mechanism of transfer of phosphate to ATP was a mystery until Lohmann showed that ADP was required.

Back in Madrid in 1931, Ochoa married Carmen García Cobian and went to the National Institute for Medical Research in London, which was directed by Sir Henry Dale who, as noted earlier, won the Nobel Prize with Otto Loewi in 1936 for his work on the chemical transmission of nervous impulses. Ochoa worked with Dale on the effects of the adrenal glands on muscle contraction.

Meanwhile, there were exciting developments back in Madrid, where one of the professors in the medical school had conceived of creating an institute of medical research and had raised a considerable amount of money for it. Ochoa was offered the direction of the physiology section and began work there in 1935. However, this was suspended due to the Spanish Civil War. Ochoa said that he would often encounter the bodies of assassinated persons in the streets near the medical school. So he and his wife decided to go to Heidelberg where he would work with Meyerhof again. Further trouble came because Meyerhof was Jewish, and the Nazis were on the rise. Initially, the German authorities left the Ochoas alone, but later they confiscated their passports. The Spanish embassy in Berlin then questioned why Ochoa was in Germany rather than Spain, but he managed to convince them that it was better for him to be in Germany, and their passports were returned. Ochoa's stay in Heidelberg was short because, as noted in chapter 3, in 1938 Meyerhof left for Paris and then the United States. Before leaving, he arranged through his friend, A. V. Hill, a fellowship for Ochoa to go to the Marine Biological Laboratory in Plymouth, England. As already noted, Hill had shared the 1922 Nobel Prize with Meyerhof for his work on muscle energy metabolism. Ochoa had a pleasant time in Plymouth, but his fellowship was only for six months.

He then went to work with Rudolph Peters, who was the chair of biochemistry at Oxford University. Peters had discovered glutathione and had studied the role of the vitamin thiamine in pyruvate oxidation in the brain. Peters was knighted and held the chair at Oxford for a long time (1923–1954) until he was succeeded by Hans Krebs. In his autobiography, Ochoa says he was impressed by the British knack of accomplishing very much without apparently working too hard, a trait that he fortunately found to be contagious. He also describes how Ernst Chain became so excited while describing the properties of penicillin that he literally jumped up on his chair. Chain would win the Nobel Prize in 1945, together with Alexander Fleming and Howard Florey, for his work on this miracle drug. Unfortunately, World War II cut short Ochoa's tenure at Oxford, which he described as a small university town with a deep cultural tradition that was ideal for intellectual accomplishment. Ochoa wrote to Carl Cori for a position in his laboratory and was elated when he was accepted.

Research with the Coris and at New York University

Ochoa said that he and his wife sailed for America not without sadness, but full of hope and expectations. He found the Cori laboratory an exciting place where enzymes were the focus, especially phosphorylase, and his coworkers were stimulating (Figure 10). The project proposed by Carl Cori seemed relatively straightforward, but turned out to be frustrating. It was known that fructose could be converted to glucose, and it seemed that fructose would first be converted to fructose-6-phosphate, which would then be isomerized to glucose-6-phosphate, which would then be dephosphorylated to glucose. But the initial experiments showed that a different compound was the first product, namely fructose-1-phosphate, and Ochoa could never show its conversion to fructose-6-phosphate or glucose-1-phosphate. Despite his brilliance as a biochemist, he left the laboratory feeling discouraged. The pathway was later elucidated by Henri-Gery Hers in Leuven, Belgium, and turned out to be rather convoluted, involving cleavage of fructose-1-phosphate by aldolase (discovered by Meyerhof and Lohmann) to glyceraldehyde and dihydroxyacetone phosphate, which was then isomerized to glyceraldehyde-3-phosphate. These two compounds were then converted by aldolase to fructose-1,6-bisphosphate, which then entered the gluconeogenic pathway to be converted to glucose.

In 1942, Ochoa was recruited to New York University (NYU) as a research associate in the Department of Medicine. He was later made an assistant professor of biochemistry and in 1946 was appointed to the chair of pharmacology—a meteoric rise unusual even in U.S. medical schools. At first, he was accommodated in Bellevue Hospital, but his space was needed for others, and Isidor Greenwald, an expert on thyroid diseases, kindly gave him space in his laboratory in the NYU Medical School. When Ochoa was offered the chair of pharmacology, he wasn't sure he wanted the job, but felt that it would allow him to expand his research efforts. In addition, it meant he could move into some beautiful new laboratories in an older building. In 1954, he moved to the chair of biochemistry in a brand new building.

At NYU, Ochoa was blessed with an outstanding graduate student, Alan Mehler, and two very gifted postdoctoral students, Santiago Grisolia and Arthur Kornberg (Kornberg is the subject of a later chapter). Mehler went to Chicago and then to the National Institutes of Health (NIH), where he worked with Herbert Tabor. Mehler worked mainly on transfer RNA (whose role in protein synthesis will be described later) and became a professor of biochemistry at Howard University. Grisolia continued to be active in the areas of glycolytic enzymes and those involved in the urea cycle. After taking positions in several U.S. medical schools, Grisolia returned to Spain where he became a prominent figure in Spanish science and in UNESCO.

Ochoa continued his work on oxidative phosphorylation, which he had begun at Oxford. Like Kalckar, he found this rather unproductive—part of the reason was that the process occurred in mitochondria, as recognized by American biochemists Albert Lehninger and Eugene Kennedy, and therefore couldn't be studied in soluble cell extracts. Lehninger received his PhD from the University of Wisconsin and was initially interested in fatty acid oxidation in liver and found that the reaction was somehow coupled to phosphorylation. He took a faculty position at the University of Chicago and focused his research on oxidative phosphorylation. As noted above, Lehninger made the major discovery that the process occurred in the mitochondria and he also showed with Morris Friedkin that the electron transport from NADH to O_2 was the direct energy source for oxidative phosphorylation. He also wrote an excellent textbook on biochemistry and authored two other textbooks.

Ochoa reasoned that the mechanism of oxidative phosphorylation was not likely to be solved unless the enzymes involved were understood, in particular those of the citric acid cycle. Warburg and David Keilin had shown that the transfer of hydrogen and electrons from oxidizable substrates to oxygen involved pyridine nucleotides (now called nicotinamide adenine dinucleotides), flavoproteins, and cytochromes, but little was known about the enzymes of the citric acid cycle involved in the oxidations. The spectacular career of Warburg has already been described. Keilin was born in Moscow of Polish parentage, and, after studying medicine and biology in Liège and Paris, he went to the University of Cambridge to study parasitology. While studying hemoglobin in a horse intestinal parasite, Keilin came upon cytochrome, which he recognized as a respiratory pigment. This discovery began studies of what is now known as the electron transfer chain, which is an essential component of cellular respiration.

In his studies of the citric acid cycle, Ochoa started with isocitrate dehydrogenase, which converts isocitrate to α-ketoglutarate and CO_2 with NADP as a cofactor. He confirmed the proposal that oxalosuccinate was an intermediate in the reaction. He then tested to see if the reaction was reversible, that is, if it could result in CO_2 fixation and NADPH oxidation. His friend, the Austrian biochemist Efraim Racker, didn't think it was so crazy, so he went ahead and watched the spectrophotometer for evidence of the oxidation. As described in his autobiography, when he saw it, he rushed out of the room yelling, "come and watch this," but nobody came because in his excitement, he had forgotten it was past 9:00 p.m. Racker was born in Poland, but grew up in Vienna. When Hitler marched into Austria in 1938, he left for Britain and then the United States. He spent time at the University of Minnesota, NYU, and Yale before taking a position at the Public Health Research Institute of the city of New York. The focus of his research was oxidative phosphorylation, and he is credited with the famous maxim: "Don't waste clean thoughts on dirty enzymes."

Another reversible enzyme that Ochoa studied with Alan Mehler was malic enzyme. This catalyzes the decarboxylation of oxalacetate to pyruvate and CO_2 and also its reduction to malate by NADPH. This led to the proposal that the enzyme had two active centers. In 1948, Ochoa was joined by Joe Stern, who had been a student of Hans Krebs. Mehler went to NIH to work with Herbert Tabor on studies of histidine metabolism, and then went to the Biochemistry Department of the Medical College of Wisconsin in Milwaukee where he worked on the synthetases that load amino acids onto their specific tRNAs. Mehler concluded his career at Howard University College of Medicine in Washington, DC, and Stern went on to Western Reserve University in Cleveland, Ohio, where he studied ketone body metabolism.

With Stern, Ochoa began studies of the most elusive enzyme of the citric acid cycle viz. condensing enzyme, which starts the cycle by condensing oxaloacetate with "active acetate" to form citrate. The enzyme was crystallized, and the reaction was shown to require coenzyme A (CoA). In experiments with Feodor (Fitzi) Lynen, who discovered acetyl-CoA, the condensing enzyme was shown to reversibly convert acetyl-CoA and oxaloacetate to CoA and citrate. Lynen, who would share the Nobel Prize in 1964 with Konrad Bloch for their work on cholesterol and fatty acid metabolism, was a true Bavarian with a colorful character and an ebullient personality. He had a serious limp from a skiing accident, but this did not prevent him from fearlessly attacking the slopes.

In the course of Ochoa's work, other enzymes were discovered. Stern and Minor Coon found a transferase in muscle and heart that catalyzed the transfer of CoA from succinyl-CoA to acetoacetate. Since the tissues do not contain an enzyme that converts acetoacetate directly to acetoacetyl-CoA, the transferase is important during starvation since it allows the utilization of ketone bodies as an energy source. Coon later achieved fame for the purification and characterization of the enzyme NADPH-cytochrome P450 reductase, multiple forms of which are important in metabolic transformations of drugs, steroid hormones, and carcinogens. Another enzyme of significance studied by Ochoa's group is propionyl-CoA carboxylase, which converts the short-chain propionyl-CoA to methylmalonyl-CoA, which is then converted to succinyl-CoA, which can enter the citric acid cycle. This enzyme is of importance in the metabolism of propionic acid and propionyl-CoA, which are produced during the breakdown of certain amino acids and during the digestion of grass by ruminants.

Polynucleotide Phosphorylase and the Genetic Code

In 1954, Ochoa returned to the study of oxidative phosphorylation, but it turned out that his major discovery was not in that field. In experiments involving Marianne Grunberg-Manago from Paris and Irwin Rose from Yale, in which ^{32}P was used in a search for enzymes converting ADP to ATP, it was found that an enzyme from *Azobacter vinelandii* incorporated the isotope into ADP—not into ATP as expected. Furthermore, it labeled other nucleoside diphosphates (UDP, CDP, GDP). The enzyme was also found to hydrolyze the diphosphates. This seemed rather uninteresting until it was found that the enzyme could convert ADP to a large polymer of adenosine phosphate residues. This was also true for the other nucleoside diphosphates, and the enzyme was named polynucleotide phosphorylase. A paper describing this was published in 1955 despite receiving adverse criticism from a reviewer. However, the description of the cell-free synthesis of RNA-like molecules aroused much interest at meetings. The only disappointment was the realization that the enzyme was unlikely to be responsible for the synthesis of RNA in the cell since it did not need a DNA template. However, it proved immensely useful in unraveling the mechanism of protein synthesis. This was because it made synthetic polynucleotides readily available for research on the genetic code. Ochoa considered it as akin to the "Rosetta stone!"

Marianne Grunberg-Manago played a key role in the discovery of polynucleotide phosphorylase. She was born in 1921 into a family of artists in Petrograd (now St. Petersburg) in the former Soviet Union. Her family emigrated to France where she studied biochemistry. After her stay in the Ochoa laboratory, she returned to Paris and rose to prominence in European science. In 1955, she was named president of the French Academy of Sciences—the first woman to attain this distinction. She was also the first female president of the International Union of Biochemistry.

The concept of messenger RNA (mRNA) as the template for protein synthesis had been promulgated in 1960. Ochoa suggested that the system might respond to synthetic polynucleotides and would incorporate different amino acids, depending on their base composition. In other words, it might provide a means to decipher the genetic code. This brilliant conception proved to be the case. When the work was starting, Marshall Nirenberg (Nobel Prize 1968), using synthetic RNA polymers, reported that a cell-free bacterial system translated polynucleotides containing uracil into polyphenylalanine. Ochoa's group confirmed this and showed that other polynucleotides made other amino acid polymers. Ochoa said they ran a close race with Nirenberg, but their results agreed perfectly. It didn't take long for both groups to identify the base composition of the triplets coding for all the amino acids and thus decipher the genetic code, and to discover that it is degenerate, that is, in many cases, several triplets code for the same amino acid. The next phase of the work was accomplished by H. Gobind Khorana, who shared the Nobel Prize with Nirenberg. This involved showing that trinucleotides of specific sequence (codons) in the mRNAs promoted the binding of specific transfer RNAs (tRNAs). Due to the action of specific enzymes discovered by Paul Berg that couple each amino acid to its appropriate tRNA(s) and the specific binding of the aminoacyl-tRNAs to complementary anticodons in mRNA, polypeptides of specific amino acid sequence could be synthesized. In fact, Khorana used the system to synthesize polypeptides with specific amino acid sequences in a cell-free system. Further work by Ochoa in this area using polynucleotide phosphorylase demonstrated that mRNA was read from the 5' end to the 3' end. This was contrary to expectation since it was thought that polypeptides were synthesized from the amino terminus to the carboxyl terminus.

The final phase of Ochoa's work at NYU focused on factors required for the initiation of polypeptide chains. Several of these initiation factors were identified, and it was found that the prokaryotic forms had a single subunit, whereas those in eukaryotes could have ten or more. In 1974, Ochoa retired from the chair of biochemistry and joined the Roche Institute of Molecular Biology in Nutley, New Jersey, directed by Sidney Udenfriend, who had been a biochemistry graduate student at NYU. At the Institute Ochoa also interacted with Efraim Racker, another associate from NYU. Unfortunately, Hoffmann-La Roche later closed the Institute. Ochoa describes that, while in New York, he attended the legendary Enzyme Club, whose meetings were held at the Columbia Faculty Club and then the Rockefeller Institute. Presentations of research highlights (no slides allowed) were preceded by cocktails and dinner.

Ochoa served on many important biochemical committees and was president of the International Union of Biochemistry from 1961 to 1967. During this time, he had to deal with the sensitive issue of whether to admit members of the Taiwan society versus those from the People's Republic of China. Ultimately, both were admitted. Responding to entreaties from Spanish colleagues, Ochoa and his wife returned to Madrid in 1985, where he was appointed professor of biology at the Autonomous University there. A new research center was planned in his honor and finally built as the Severo Ochoa Molecular Biology Center. Ochoa never recovered from the death of his wife of fifty-four years and died in Madrid on November 1, 1993.

In his recollection, Ochoa describes some of the people he recruited or interacted with at NYU. It was his philosophy to impart a culture in biology to the medical students. Consequently, they were taught biochemistry at almost the level of a graduate course and given experiments in the newly emerging areas of biochemistry. The faculty attended each other's lectures, and people from other departments were often present—all of this kept the faculty on their toes! He deliberately kept his departments small, and research problems were discussed over coffee.

Faculty Recruited by Ochoa to New York University

Some of Ochoa's key faculty included Sarah Ratner, who was notable for her classic work on the enzymes of the urea cycle discovered by Hans Krebs. Ratner was born in New York City and entered Cornell University, where she was the only female student in many of her classes. She took a job at a Long Island hospital, taking evening classes at Columbia University, and was accepted as a graduate student in the College of Physicians and Surgeons. She encountered gender discrimination when she looked for a postdoctoral position, but ultimately Rudolf Schoenheimer accepted her. As will be described later, Schoenheimer also accepted Mildred Cohn under similar circumstances. Working with Schoenheimer and David Rittenberg, Ratner utilized ^{15}N to begin classic studies of the synthesis of urea.

Schoenheimer was a member of the astonishingly large group of European Jews who enriched American science in the 1930s due to the policies of Hitler. Rudolf Schoenheimer was born and educated in Berlin and, after three years in Leipzig, worked in Freiburg studying the role of cholesterol in atherosclerosis. He left Germany in 1933 for New York City and was given a position at Columbia University by Hans Clarke, who was chairman of biological chemistry there. Clarke is notable because he provided positions and support for other scientists escaping Nazi Germany (e.g., Karl Meyer, Erwin Chargaff, Zacharias Dische, Heinrich Waelsch, and Erwin Brand), who would make great contributions to biochemistry. Schoenheimer and Rittenberg initiated a whole new approach to the study of intermediary metabolism in intact animals through the procedures of administering D_2O (heavy water) or compounds tagged with deuterium or other stable isotopes and following their conversions through metabolic pathways. Rittenberg was born, raised, and educated in New York City and spent almost all his life there. He got his PhD under Harold Urey at Columbia University and, after joining Hans Clarke's department, began his fruitful collaboration with Schoenheimer. They needed a mass spectrometer for their experiments, but one was not available commercially, so Rittenberg built his own.

Another member of Ochoa's department at NYU was Guilio Cantoni, who discovered "active methionine" (S-adenosylmethionine), which donates methyl groups to a variety of proteins. Cantoni was born in Milan and received his MD from the University of Milan in 1939, but had to leave Italy with his family to escape the anti-Semitic laws. Sir Henry Dale arranged a fellowship for him at the National Institute for Medical Research in London. With the outbreak of war, the Institute was moved out of London, and Cantoni spent a year at Oxford University. He decided to emigrate to the United States, but when he and his family were about to board a ship in Liverpool, he was arrested as an enemy alien. This was because Italy had just joined Germany and had declared war on England. Cantoni was shipped across the Atlantic and interned in Canada as a prisoner of war. Eventually, the British government recognized his status as a refugee, and he finally joined his mother and sister in New York in 1941. After a job at the University of Michigan, he joined Ochoa's department at NYU.

Another outstanding member of the department was M. Daniel (Dan) Lane, who obtained his PhD from the University of Illinois. He purified propionyl-CoA carboxylase and determined that it required biotin for activity and that the biotin was linked to a lysine residue in the enzyme. Working with Fitzi Lynen in Munich, Lane went on to purify the enzyme that loaded biotin on to propionyl-CoA carboxylase. After joining Ochoa's department, he isolated another biotin-containing enzyme, namely acetyl-CoA carboxylase, which converts acetyl-CoA to malonyl-CoA. Its importance is that it catalyzes the first step in fatty acid synthesis. Lane left New York in 1970 to become a professor at Johns Hopkins Medical School, where he continues to make fundamental observations about acetyl-CoA carboxylase.

Arthur Kornberg described Ochoa as a courtly, charming El Greco-like figure who, with his wife Carmen, was a most gracious host. They especially enjoyed music, fine food, travel, and good company. Kornberg recalled an incident when he was involved in the purification of malic enzyme. Ochoa's devoted assistant had worked for several weeks purifying the enzyme from several hundred pigeon livers, and Ochoa and Kornberg were pouring the final fraction into a measuring cylinder when Kornberg overturned a series of bottles, which in turn tipped the contents of the cylinder onto the floor, thus apparently ruining the entire enzyme preparation. Despite reassurance from Ochoa, Kornberg was greatly upset. He took the subway home and recalls that Ochoa called him several times there because he was concerned about his safety. It turned out that Kornberg had saved an earlier fraction and kept it in a freezer. Kornberg noticed that this fraction had become turbid. And so he assayed it and found that, in fact, this fraction contained most of the enzyme activity. When he discovered this, he shrieked, "Holy Toledo," much to the amusement of Ochoa.

Dan Lane also describes Ochoa as having a princely presence because of his carriage, tall stature, and silver hair. Despite the fact that his presence inspired awe, he had a warm personality and a genuine concern for others. Lane describes his experience giving his first lectures to medical students at NYU. He knew his lectures weren't particularly good, but Ochoa put his hand on his shoulder and said, "That was an excellent lecture, Dan." This was an example of how he encouraged his faculty and students.

He had a particular celebration for his seventy-fifth birthday in 1975, with guests being those scientists he most respected worldwide. Symposia and celebratory dinners were held in both Barcelona and Madrid. There was even a visit to Salvador Dalí in his museum in Figueres. Kornberg described it as a party, the likes of which has not been seen in scientific circles before or since. Apart from the Nobel Prize, Ochoa received many distinctions, including 3 honorary doctorates, 120 medals, 219 diplomas, and many other decorations. Grisolia states that Ochoa's mind remained perfectly clear until the day of his death.

References

Bentley, R. 2003. "Sarah Ratner June 9, 1903–July 28, 1999" *Biogr. Mem. Natl. Acad. Sci. U.S.A.* 82: 220–41.
Cantoni, G. L. 2000. *From Milano to New York by Way of Hell: Fascism and the Odyssey of a Young Italian Jew*. Lincoln, NE: Writers Club Press.
Flavin, M., P. J. Ortiz, and S. Ochoa. 1955. "Metabolism of Propionic Acid in Animal Tissues." *Nature* 176: 823–26.
Grunberg-Monago, M., P. J. Ortiz, and S. Ochoa. 1955. "Enzymatic Synthesis of Nucleic Acid like Polynucleotides." *Science* 122: 907–10.
Hers, H.-G. 1983. "From Fructose to Fructose-2,6-bisphosphate with a Detour through Lysosomes and Glycogen" *Comprehensive Biochemistry*. Vol. 35. Edited by G. Semenza. 71–101. Zurich: Elsevier Science.
Kornberg, A. 2001. "Remembering Our Teachers." *J. Biol. Chem.* 276: 3–11.
Lane, M. D. 2004. "The Biotin Connection: Severo Ochoa, Harland Wood and Feodor Lynen." *J. Biol. Chem.* 279: 39187–94.
Lane, M. D., and P. Talalay. 1986. "Albert Lester Lehninger 1917–1886." *J. Membr. Biol.* 91: 194–97.
Langen, P., and F. Hucho. 2008. "Karl Lohmann and the Discovery of ATP." *Angew. Chem. Int. Ed.* 47: 1824–27.
Lipmann, F. 1974. "Reminisces of Emden's Formulation of the Embden-Myerhof Cycle." *Mol. Cell. Biochem.* 6: 171–75.
Notice Biographique de Marianne Grunberg-Monago, Membre de l'Académie des Sciences Institut de France.
Ochoa, S. 1980. "The Pursuit of a Hobby." *Ann. Rev. Biochem.* 49: 1–30.
Slater, E. C. 2003. "Keilin, cytochrome and the Respiratory Chain." *J. Biol. Chem.* 278: 16455–61.
Smith, M. A., M. Salas, W. M. Stanley Jr., A. J. Wahba, and S. Ochoa. 1966. "Direction of Reading of Genetic Message. II." *Proc. Natl. Acad. Sci. U.S.A.* 55: 141–47.
Speyer, J. F., P. Lengyel, C. Basilio, and S. Ochoa. 1962. "Synthetic Polynucleotides and the Amino Acid Code. II." *Proc. Natl. Acad. Sci. U.S.A.* 48: 63–68.
Stanley, W. M. Jr., M. Salas, A. J. Wahba, and S. Ochoa. 1966. "Translation of the Genetic Message: Factors Involved in the Initiation of Protein Synthesis." *Proc. Natl. Acad. Sci. U.S.A.* 56: 290–95.
Stern, J. R., S. Ochoa, and F. Lynen. 1952. "Enzymatic Synthesis of Citric Acid. V. Reaction of Coenzyme A." *J. Biol. Chem.* 198: 313–21.
Stern, J. R., M. J. Coon, A. del Campillo, and M. C. Schneider. 1956. "Enzymes of Fatty Acid Metabolism. IV. Preparation and Properties of Coenzyme A Transferase." *J. Biol. Chem.* 221: 15–31.

5

Move to Enzymology and Work of Arda Green

Studies with Phosphorylase

The Coris didn't stop at the identification of glucose-1-phosphate, but searched for the enzymes responsible for the formation of this compound and for its further metabolism. The first enzyme they studied was phosphoglucomutase, which converts glucose-1-phosphate to glucose-6-phosphate. They had observed this reaction previously, because it had complicated their studies of glucose-1-phosphate formation from glycogen. After the enzyme was purified by Victor Najjar, it was further characterized by several studies of the Cori group and found to be accelerated by Mg^{2+} and some other divalent ions. In fact, the virtual absence of Mg^{2+} ions in the extracts used in their earlier experiments enabled them to isolate glucose-1-phosphate. When this enzyme was studied in detail, it was found that the rate of the reaction was very variable, and it was suspected that this was due to the presence of an activator. The nature of this activator remained obscure until Luis Leloir, whose career is described below in chapter 6, suggested that it might be glucose-1,6-bisphosphate. This was confirmed by the Cori group as was the mechanism of action proposed for the enzyme by Leloir.

A more important enzyme from a regulatory viewpoint was the one that broke down glycogen in response to epinephrine. The Polish biochemist Jacob Parnas had termed the reaction phosphorolysis because it involved inorganic phosphate. So the enzyme responsible was named phosphorylase. The study of this enzyme would lead to great advances in understanding how hormones act and how the body responds to exercise. Interestingly, phosphorylase was found to synthesize as well as break down glycogen, and it was found that both reactions were stimulated by AMP. As described below, another sequence of enzymes was discovered by Luis Leloir that synthesized glycogen, and this is now considered the major pathway by which glycogen and starch are synthesized. Leloir spent some time in the Cori laboratory in 1943. The Coris made a detailed study of the synthesis of glycogen from glucose-1-phosphate as catalyzed by phosphorylase, although the physiological significance of the reaction is now disputed. They studied the kinetics of the reaction and showed that they obeyed the classic Michaelis-Menten equation described for enzymes by the German biochemist Leonor Michaelis working with a Canadian student, Maud Menten, in 1913. This defines the relationship between the reaction rate and the substrate concentration and has been described as the foundation of modern enzymology. The Coris noticed that there was a lag period when the synthesis of glycogen from glucose-1-phosphate was studied in tissues other than liver and realized that small amounts of glycogen were required to prime the reaction. Later, when they obtained phosphorylase in a pure form, they realized that another enzyme was required for glycogen synthesis. This was because phosphorylase could only form 1:4 linkages between glucose moieties, whereas glycogen had 1:6 linkages at its branch points. They therefore proposed that another enzyme was involved. As described below, this proposal proved to be correct.

In a testament to Gerty Cori's and Arda Green's experimental skills, phosphorylase was crystallized from muscle. To put this achievement in perspective, the crystallization of an enzyme is no mean feat. It requires purification by different procedures that must be conducted at low temperatures to minimize degradation of the enzyme. Because there are often considerable losses during isolation of the enzyme, large amounts of starting material are needed, and crystallization requires the isolation of milligram quantities of the enzyme in pure form. Then a variety of crystallization conditions must be tested before success, which is by no means guaranteed. To facilitate the purification of enzymes, the Coris later used part of their Nobel Prize money to construct a small cold room in Gerty's laboratory. The isolation of large amounts of pure phosphorylase permitted its very detailed characterization as depicted in four classic papers in the *Journal of Biological Chemistry* in 1943. These papers described the physical properties of the enzyme and, most importantly, they showed that it occurred in two forms. One of these (phosphorylase *a*) was active in the absence of AMP, whereas the other (phosphorylase *b*) required AMP for activity. It is now known that the effect of AMP represented the first example of allosteric control of an enzyme. The concept of allostery was later recognized by the French investigator Jacques Monod, together with Jeffries Wyman and Jean-Pierre Changeux. Monod won the Nobel Prize in 1965, together with François Jacob and André Lwoff, but for work on the regulation of genes. As illustrated in the case of the Coris, Severo Ochoa, Monod and his associates, and others in this book, investigators can make fundamental discoveries in more than one area of research.

The discovery that phosphorylase existed in two forms of different activities had important repercussions with respect to elucidating the mechanism of action of many hormones and neurotransmitters. It also led to the discovery of other major enzymes involved in the regulation of cellular processes. The first of these enzymes was the one that inactivated phosphorylase by converting it from the *a* form to the *b* form. The Coris postulated that the inactivation was due to the removal of a prosthetic group (AMP), but in a rare example, they were wrong, because phosphorylase *a* does not contain this nucleotide. Later work in their laboratory by Earl Sutherland, whose career is described in chapter 7, and by Edwin Krebs and Edmond Fischer at the University of Washington, showed that the conversion of phosphorylase *b* to *a* involved enzyme-catalyzed phosphorylation, and the reverse reaction involved enzyme-catalyzed dephosphorylation. The two enzymes responsible for these changes were respectively called phosphorylase kinase and phosphorylase phosphatase. The phosphate donor for the kinase was found to be ATP, as expected, but there was an unexpected complication, which will be explained later. Phosphorylase kinase represented the first example of a large family of enzymes termed protein kinases that phosphorylate and alter the functions of proteins. This work—plus the identification of another enzyme, cyclic AMP-dependent protein kinase (PKA), which activated phosphorylase kinase—resulted in Krebs and Fischer being awarded the Nobel Prize in 1992. As will be described in a later chapter, Krebs also spent some time in the Cori laboratory.

Arda Alden Green—Protein Chemist

Arda Green played a crucial role in the Coris' studies on phosphorylase. Compared with the extensive biographical material available for Herman Kalckar and Severo Ochoa, information on Green is sparse, and I am indebted to her obituary written by Sidney Colowick. She was born in Prospect, Pennsylvania, on May 7, 1899, and obtained her AB from the University of California, Berkeley. She received highest honors in chemistry and honors in philosophy and, after one year of graduate study, decided to study medicine. During her first two years, she became acquainted with Herbert Evans, an endocrinologist at Berkeley who was notable for isolating and identifying several hormones. He encouraged her to interrupt her studies and spend a year at Harvard in the laboratory of Edwin Cohn, who was one of the few protein chemists in the United States. Green moved to Johns Hopkins and obtained her MD in 1927. While there, she did some research with Leonor Michaelis, whose famous equation with Maude Menten is alluded to above. This involved studies of the conductivity of electrolytes within membranes.

Green returned to Cohn's laboratory for the period 1927–1929 and carried out classic studies on the effects of pH on the binding of O_2 to hemoglobin—these formed the basis for some of Linus Pauling's views of the structure of this protein. She remained at Harvard for the next twelve years, first as a research fellow in the laboratory of Cohn and of Lawrence J. Henderson (head of physical chemistry at Harvard and notable for the Henderson-Hasselbach equation for pH buffers), and then as a research associate in pediatrics. However, she remained in contact with Cohn and played a significant role in his development of procedures for isolating plasma proteins. This work was of major importance during

World War II and later, when large amounts of albumin, γ-globulins, fibrinogen, and thrombin were needed for the treatment of military personnel. Cohn was a demanding mentor and supervisor. One of his trainees, John Edsall (later editor of the *Journal of Biological Chemistry*), depicted Cohn's personality in unflattering terms: "In applying pressure to get things done he was often imperious and demanding, sometimes rude to the point of insult. Certainly he made many enemies by his extreme forcefulness and sometimes by an aggressiveness which went beyond obvious necessity." Nevertheless, Cohn's contributions to the war effort were significant and were rewarded in 1948 by a Medal for Merit from the U.S. government.

In addition to her work on the isolation of plasma proteins, Green engaged in a thorough study of the solubility of hemoglobin under different conditions of pH and ionic strength. From 1933 to 1937, she published studies of the combination of hemoglobin with CO_2 and CO and of the interaction of globulins with Ca^{2+} ions. During her time in Boston, she also served as a tutor in biochemical sciences at Radcliffe College. She spent the next phase of her career (1941–1945) in the Cori laboratory, where she held some nontenure positions. She utilized her skills in protein purification to isolate and crystallize phosphorylase, and the studies of this enzyme are considered to be a major factor in the Coris receiving the Nobel Prize. She was later involved in the crystallization of aldolase, and Colowick believed that her presence in the laboratory had a tremendous influence on the development of enzymology in the United States. This was because she demonstrated the importance of having pure enzymes for studying reactions.

After her stay in St. Louis, she went to the Cleveland Clinic to work in the laboratory of Irvine Page, a preeminent investigator in the field of hypertension. Her project was to isolate a vasoconstrictor substance from plasma. This she did in collaboration with Page and M. M. Rapport. The compound was named serotonin, and its structure was determined by Rapport. Now we know it to be an important neurotransmitter involved in many central nervous functions and to be implicated in many neuropsychiatric disorders, especially bipolar disease. It also plays a role in gastrointestinal motility. While in Cleveland, Green also worked on the purification of proteins involved in renal hypertension, for example, angiotensinogen and angiotensin. Cleavage of angiotensinogen by the enzyme renin yields angiotensin I, which is then converted to the active compound angiotensin II, and this, through its effects on blood vessels and the adrenal cortex (aldosterone secretion) raises the blood pressure. Green's work in this area was obviously of great importance in understanding the disease of hypertension. Despite her skills in protein purification, she was, however, unable to isolate renin.

In 1953, she moved to Johns Hopkins to become yet another researcher who is mentioned in this book and who worked in the laboratory of W. D. McElroy at the McCollum-Pratt Institute. There, she joined him on his classic work on the luciferin/luciferase system for light generation in fireflies. Her major accomplishment was to crystallize luciferase, and this led to elucidation of the enzyme mechanism. Her next project was the purification of the enzyme responsible for bacterial luminescence. Sadly, she died on January 22, 1958, of a crippling disease before this project could be completed.

Colowick notes that she was importantly involved in the training of graduate students at Johns Hopkins. She not only gave authoritative lectures on proteins, but also established warm personal relationships with the students. Arda Green did not always receive the recognition she deserved. This was partly due to the stature of the men she worked with (Cori, Page, McElroy), but also reflected the position of women in the field of science at that time, a theme that will merge in the section devoted to Mildred Cohn. Green was clearly gifted in the techniques of protein purification and crystallization, which some attributed to her "magic touch." Others ascribed her prowess in crystallization to the ash that fell from her constant cigarette. She personally attributed it to "horse sense." Colowick notes that her life was not completely dominated by science, and she brought the same vigor, enthusiasm, and thoroughness to cooking, dressmaking, music, and entertaining. She made sure that any unattached people in the laboratory would not go without a Thanksgiving dinner. Her genuine concern for others was reflected not just in words but also in practical, helpful gestures.

References

Colowick, S. P. 1958. "Arda Alden Green." *Science* 128: 519–21.
Cori, C. F., G. T. Cori, and A. A. Green. 1943. "Crystalline Muscle Phosphorylase. III. Kinetics." *J. Biol. Chem.* 151: 39–55.
Cori, G. T., S. P. Colowick, and C. F. Cori. 1938. "The Enzymatic Conversion of Glucose-1-Phosphoric Ester to 6-Ester in Tissue Extracts. *Biol. Chem.* 124: 543–55.
Cori, G. T., and C. F. Cori. 1940. "The Kinetics of the Synthesis of Glycogen from Glucose-1-Phosphate." *J. Biol. Chem.* 135: 733–56.
Cori, G. T., and C. F. Cori. 1943. "Crystalline Phosphorylase. IV. Formation of Glycogen." *J. Biol. Chem.* 151: 57–63.
Cori, G. T., and C. F. Cori. 1945. "The Enymatic Conversion of Phosphoryase *a* to *b*." *J. Biol. Chem.* 158: 321–32.
Cori, G. T., and A. A. Green. 1943. "Crystalline Muscle Phosphorylase II Prosthetic Group." *J. Biol. Chem.* 151: 31–38.
Edsall, J. T. 1961. "Edwin Joseph Cohn." *Biogr. Mem. Natl. Acad. Sci. U.S.A.* 20: 47–72.
Green, A. A., and G. T. Cori. 1943. "Crystalline Muscle Phosphorylase: I Preparation, Properties, and Molecular Weight." *J. Biol. Chem.* 151: 21–29.
Green, A. A., G. T. Cori, and C. F. Cori. 1942. "Crystalline Muscle Phosphorylase." *J. Biol. Chem.* 142: 447–48.
Leloir, L. F., C. E. Trucco, A. Cardini, A. Paladini, and R. Caputto. 1948. "The Coenzyme of Phosphoglucomutase." *Arch. Biochem.* 19: 339–40.
Monod, J., J. Wyman, and J. P. Changeux. 1965. "On the Nature of Allosteric Transitions: A Plausible Model." *J. Mol. Biol.* 12: 88–118.
Najjar, V. A. 1948. "The Isolation and Properties of Phosphoglucomutase." *J. Biol. Chem.* 175: 281–90.

Randle, P. J. 1986. "Carl Ferdinand Cori." *Biogr. Mems. Fell. R. Soc.* 32: 66–95.
Sutherland, E. W., T. Z. Posternak, and C. F. Cori. 1949. "The Mechanism of Action of Phosphoglucomutase." *J. Biol. Chem.* 179: 501–02.
Sutherland, E. W., M. Cohn, T. Posternak, and C. F. Cori. 1949. "The Mechanism of the Phosphoglucomutase Reaction." *J. Biol. Chem.* 180: 1285–95.
Taylor, J. F., A, A. Green, and G. T. Cori. 1948. "Crystalline Aldolase." *J. Biol. Chem.* 173: 591–604.

6

Luis Leloir

One of Argentina's Greatest Scientists

Education in Buenos Aires and Influence of Bernardo Houssay

Luis Frederico Leloir spent only a short time in the Cori laboratory, but interacted with both Coris, Sidney Colowick, Arda Green, and others with whom he met daily (Figure 11). He regarded the laboratory highly and was attracted by their reports on the crystallization and characterization of phosphorylase. Leloir was born in Paris just a few blocks away from the Arc de Triomphe on September 6, 1906. His parents were from Argentina and took him to Buenos Aires when he was two years old. His father was educated as a lawyer, but never practiced law. Leloir grew up in a house full of books and maintained there was no specific reason he was attracted to science. He described himself as ill-suited for music, sports, politics, or law. There is a story that in the 1920s, he invented *salsa golf*. After being served prawns with the usual sauce at a lunch with friends at the Ocean Club in Mar del Plata, he thought up a combination of ketchup and mayonnaise to spice up the meal. Later, when financial difficulties plagued his research, he would joke, "If only I had patented that sauce, we'd have a lot more money now."

After completing his primary and secondary education at the Escuela General San Martin and Colegio Lacordaire with unspectacular grades, he entered medical school at the University of Buenos Aires, where he was reputed to have required four attempts to pass anatomy, but graduated in 1932 and worked at the University Hospital for two years. He was discouraged by how little the doctors could do for patients, apart from surgery, digitalis, and a few other remedies. In his memoir, he tells how he felt that medical treatment was little better than that exemplified by the French story, in which the doctor ordered: "Today we shall bleed all those on the left side of the ward and give a purgative to all those on the right side." His experiences in the practice of medicine led him to join efforts with those doing research to advance medical knowledge.

The most active laboratory in the medical school was that directed by the professor of physiology, Bernardo Houssay, who, as noted earlier, shared the Nobel Prize with the Coris in 1947 for his demonstration of the role of the anterior pituitary gland in the regulation of carbohydrate metabolism, particularly in diabetes. Houssay suggested that Leloir could do his thesis work under his direction. Out of several topics, Leloir chose to study the role of the adrenal glands in regulating carbohydrate metabolism. Because his ignorance of chemistry was abysmal, he took courses at the university. Houssay helped him a lot, including performing most of the adrenalectomies, and the two began a lifetime relationship. Houssay made daily rounds of the laboratories, leaving messages on tiny pieces of paper, and this experience engendered in Leloir a lifelong concern about economy. Leloir's thesis won the Faculty Prize for the best thesis. He began to spend more time in the laboratory and less time at the hospital. He didn't have to earn a living as a physician because his great-grandparents, Basque immigrants from Spain and France, had bought a large property, *El Tuyú*, in the early days of the settlement of Argentina, and the profits from this allowed him to devote himself to research.

When his thesis was finished, Houssay advised Leloir to work abroad, and he decided that a good place to do further research would be the biochemical laboratory at Cambridge, directed by Sir Frederick Gowland Hopkins, who discovered glutathione and shared the 1929 Nobel Prize with Christiaan Eijkman for their discovery of certain vitamins. Leloir describes Cambridge at the peak of its glory with Ernest Rutherford and Paul Dirac in physics and Hopkins and David Keilin in biochemistry and parasitology. Leloir began work with Malcolm Dixon on succinate dehydrogenase, a key enzyme in the citric acid cycle. Dixon was an expert on enzymes and enzyme kinetics. Leloir also worked with Norman Edson on ketogenesis using liver slices. Edson—who greatly admired Hans Krebs, with whom he had worked—returned to his native New Zealand and became the founding professor of biochemistry at the University of Otago Medical School in 1949. He was my preceptor when I graduated PhD in 1963. With Edson's departure, Leloir worked with David Green on the ketogenic enzyme β-hydroxybutyrate dehydrogenase. Green had a strong personality and took a faculty position at Columbia University and then the University of Wisconsin. Later he became an important figure in the controversy surrounding the mechanism of oxidative phosphorylation.

After his year at Cambridge, Leloir returned to the Institute of Physiology in Buenos Aires and began working with Juan Mauricio Muñoz. Muñoz had accumulated degrees in medicine, chemistry, and dentistry in order to become the professor of physiology in the dental school, a maneuver that Leloir did not approve of. Their project involved a study of fatty acid oxidation in tissue homogenates—it had previously been believed that this could only be studied in intact tissues. After initial failures, they realized that they had to work fast and at low temperatures. To do this, they wrapped an automobile inner tube filled with a freezing mixture around the ancient centrifuge used to separate the particulate fraction from the supernatant.

The particulate fraction could oxidize the fatty acids, but required some additions (cytochrome c, fumaric acid, and ATP). Although they suspected (correctly) that mitochondria were involved, they could not prove it.

Research on Hypertension

Another project Leloir worked on while at the Institute of Physiology concerned the mechanism of hypertension caused by kidney disease. Leloir joined a group there which included Eduardo Braun Menendez, Juan Carlos Fasciolo, and Muñoz. Following the experiments of Harry Goldblatt, a pathologist at Western Reserve University, they constricted the renal arteries of dogs, which caused permanent hypertension. Grafting a damaged kidney into a normal dog produced hypertension, as did blood from a constricted kidney. Attempts to identify the substance that raised the blood pressure were initially unsuccessful. Although another pressor substance called renin had been extracted from kidneys in 1898, this was not the same as the one they had found. However, incubation of blood plasma with renin produced their pressor substance, which they called hypertensin. They suggested, correctly, that renin acted like an enzyme to release hypertensin from a precursor (hypertensinogen) in blood plasma. They also found an enzyme in tissues and blood that destroyed hypertensin, thus explaining their earlier difficulties.

At the same time, similar experiments were being carried out by Irwin Page in the laboratories of Eli Lilly in Indianapolis. Like Leloir and associates, these researchers found that renin required the presence of blood to generate a vasoconstrictor activity, which they called angiotensin. To eliminate confusion in the field, Braun Menendez and Page agreed on a Solomonic solution to call hypertensin angiotensin and the precursor angiotensinogen. Leloir's group was considerably disheartened by all this, because they could only claim a co-discovery, but not a discovery. As Leloir looked back on this incident after many years, he felt that their feelings were quite infantile. Leloir and his associates decided not to attempt to purify and determine the structure of angiotensin because their methods were too primitive. As noted in chapter 5, Arda Green was able to accomplish this working in Irwin Page's laboratory. It is now known that renin splits a leucyl-leucyl bond in angiotensinogen to yield angiotensin I, which is an inactive decapeptide. This is then acted on by angiotensin converting enzyme (ACE) to release the dipeptide histidyl-leucine and form angiotensin II, which is active. Pharmaceutical inhibitors of ACE are now widely used in the treatment of hypertension. Leloir truly enjoyed the year spent on the project because of the congenial atmosphere and the personalities of his teammates. Braun Menendez was full of energy and enthusiasm, Muñoz was original and had unique ideas, and Fasciolo told jokes and funny stories, although he was serious about his work.

Research in the United States and Return to Buenos Aires

Work in the Institute was interrupted by some disagreeable events. Houssay had innocently cosigned a letter critical of the military government, which was published in the newspapers. The government overreacted and demanded the dismissal of all the signers who held positions in state institutions. This meant that many of the best professors lost their posts, including Houssay. Most of the Institute staff resigned in protest, and work had to be continued in a private institution and organized from scratch.

Leloir thought it was a good time to leave and seek a position in the United States. The disruption in his life was ameliorated by a happy marriage to Amelie Zuberbuhler in 1943. Because this was before the jet age and commercial planes were driven by twin propellers and only flew by day, the flight to New York required several stops. Since he had made no prior arrangements, he had to start looking for a job. One place that was highly regarded was the Cori laboratory, so he traveled to St. Louis and Carl Cori kindly accepted him. He found the environment stimulating, but spent only six months there, working on the enzymatic synthesis of citric acid, one of the key reactions of the citric acid cycle where acetyl-CoA enters the cycle. To broaden his experiences, he arranged to work again with David Green, now at the College of Physicians and Surgeons of Columbia University. Leloir interacted with Sarah Ratner, Eugene Knox, and Paul Stumpf, all of whom would have distinguished careers in biochemistry. Leloir worked on the purification of two aminotransferases (enzymes that transfer amino groups from amino acids to carbon chains). Green was always full of ideas and had an impressive list of problems to be solved, but Leloir told him he didn't know how to attack any of them. He told Green about his experiments on fatty acid oxidation and his suspicion that mitochondria might be involved. Green was incredulous. Little did he realize that these organelles would become central to much of his future research.

In 1945, Leloir returned to Buenos Aires to the Institute of Physiology, where Houssay had been reinstated. Leloir endeavored to establish a small research group and recruited Ranwel Caputto, who had worked with the distinguished enzymologist Malcom Dixon at Cambridge on glyceraldehyde-3-phosphate dehydrogenase. In 1946, Houssay was approached by Jaime Campomar, a rich textile company owner who wanted to finance a research institute. Houssay suggested Leloir as a possible director. However, the faculty were shocked to learn that Houssay had been removed again from his positions as professor of physiology and director of the Institute of Physiology on the pretext he was overage. This time it was by the Peronista government—one and was one year before he got the Nobel Prize. Most of his group resigned and moved into a poorly furnished research institute with an adjacent small four-room house. The institute consisted of a room with a refrigerator and a few pipettes. Despite these bad working conditions, the researchers were young and enthusiastic and had hope in the future. They were joined by Carlos

Cardini, Naum Mittelman, and Alejandro Paladini. Campomar provided an annual contribution of 100,000 pesos, a very generous gift which was used to install the laboratory, purchase equipment, and pay some salaries. This was the beginning of the Instituto de Investigaciones Bioquimicas Fundacion Campomar (now Fundacion Instituto Leloir), where Leloir remained until his death.

Discovery of Uridine Diphosphate Glucose

The research at the new institute started with an interesting observation by Caputto. He had found that homogenates of mammary glands produced lactose when incubated with glycogen. This intrigued Leloir because at that time, the only disaccharide that had been synthesized in vitro was sucrose, and the enzyme used was bacterial. The results of early experiments were ambiguous, so he changed and instead looked at galactose utilization; he obtained evidence for an enzyme that phosphorylated galactose to galactose-1-phosphate, which was then converted to glucose-1-phosphate and then to glucose-6-phosphate. Importantly, they realized that cofactors were required for both of these reactions. They set out to find the cofactor for the second (phosphoglucomutase) reaction. Initially, they thought it was fructose-1,6-bisphosphate, an intermediate in the glycolytic pathway, but they then realized that it was glucose-1,6-bisphosphate, as described previously in chapter 4. Identifying the cofactor for the other reaction (conversion of galactose-1-phosphate to glucose-1-phosphate) was not so easy, but it led to a paradigm-shifting biochemical discovery. They found that the factor absorbed UV light with a peak at 260 nm. This was similar to that of adenine, but had some differences. The breakthrough came when Caputto arrived at the laboratory one morning with the latest copy of the *Journal of Biological Chemistry* that showed the absorption spectrum of uridine, which was identical to that of the cofactor. In addition to uridine, they found that the cofactor contained glucose and two phosphates, although their method for identifying glucose was initially poor. It was named uridine diphosphate glucose (UDPG), and its structure was later confirmed by Alexander Todd and coworkers at Cambridge. Prior to this time, sugars were known to be linked to phosphate or to other sugars, but the idea that they could be linked to a nucleotide was revolutionary.

The role of UDPG in the reaction by which galactose-1-phosphate is converted to glucose-1-phosphate is shown by the following two equations.

$$\text{galactose-1-phosphate} + \text{UDP-glucose} \leftrightarrow \text{glucose-1-phosphate} + \text{UDP-galactose}$$

$$\text{UDP-galactose} \leftrightarrow \text{UDP-glucose}$$

Herman Kalckar suggested that the first enzyme be called uridyl transferase and the second epimerase. As described in the chapter devoted to Kalckar, uridyl transferase is deficient in the disease of galactosemia.

UDPG was unique in that other known enzyme cofactors contained adenine. Leloir's group reasoned that it should have other functions and were excited to find that it was involved in the synthesis of trehalose, sucrose, and glycogen. Another novel discovery was that there was a contaminant of UDPG, which turned out to be UDP-*N*-acetylglucosamine. This is now known to be involved in the synthesis of bacterial cell walls, mucoproteins, and chitin, the most abundant protein in the animal kingdom. UDP-glucuronic acid was discovered by another group and is important in the metabolism and disposal by the liver of the bile pigment bilirubin and certain drugs. Yet another member of the family that was isolated was UDP-acetylglucosamine, which is involved in the synthesis of hyaluronic acid and chitin, which are very abundant and important glucosaminoglycans. Hyaluronic acid is found in all animal tissues and fluids, whereas chitin is the main component of the exoskeleton of arthropods.

Another compound containing guanosine instead of uridine was discovered by Enrico Cabib, a fellow who replaced Paladini, and this was shown to be GDP-mannose. By now, other groups entered the picture, and other sugar nucleotides were discovered, including TDP-glucose, ADP-glucose, GDP-glucose, and UDP-xylose. It is now known that sugar nucleotides are involved in the synthesis of a vast array of complex carbohydrates, glycoproteins, and lipopolysaccharides. These are found in animal, yeast, plant, and bacterial cells and are major constituents of cell walls and surfaces, where they are involved in communication with the environment and signal transduction. In mammalian cells, they are also involved in intracellular membrane trafficking, for example, in Golgi and lysosomes. In bacteria such as *Staphylococcus aureus*, penicillins and cephalosporins exert their bactericidal action by inhibiting a reaction involved in the synthesis of a glycoprotein required for stability of the cell wall.

In 1957, Campomar died, leaving the institute without resources. Before disbanding, they applied to NIH for a grant. At that time, NIH was making awards to foreign scientists, and they were surprised and grateful to receive a grant which allowed them to continue their research. Of particular interest to Leloir's group was the synthesis of glycogen and starch. They were able to detect the formation of glycogen from UDPG using extracts from liver and muscle. At that time, glycogen synthesis was thought to occur by a reversal of the phosphorylase reaction, although this was questioned by Earl Sutherland. The search for the enzyme responsible for the reaction was stimulated by Leloir's reading a book by Herman Niemeyer, who suggested that glycogen was formed from UDPG and not glucose-1-phosphate. The search received an impetus when an activating effect was found with glucose-6-phosphate. The enzyme, now called glycogen synthase, was then studied in detail by Leloir, Cardini, and three other associates (this is described in chapter 19, devoted to Joseph Larner). Leloir and his associates were offered space in a large house which had been a girl's school. Soon after this, the National Research Council was established, and they became associated with the University of Buenos Aires.

Synthesis of Complex Carbohydrates

Leloir and his associates expanded their research to the synthesis of sucrose and starch. Using wheat germ extracts, they showed that sucrose was synthesized from UDPG-glucose and fructose. It seemed to them the system for synthesizing starch should be similar to that for glycogen, but their initial results were negative. However, they were joined by E. Recondo, an organic chemist who embarked on a project that Leloir thought was silly. This was to synthesize nucleotides of glucose with different bases and test them with an enzyme found in starch grains. One of the first to be tested was adenosine diphosphate glucose (ADPG). This was more active than UDPG, but Leloir was not sure that the result was real until an enzyme was found that synthesized ADPG. In work by others, UDP-glucose was found to be transferred to cellulose in *Acetobacter xylinum*, whereas GDP-glucose was the donor in plants.

The research took a different turn when several groups in the United States discovered that the transfer of nucleotide sugars to certain polysaccharides involved lipid intermediates. This surprising finding came from the laboratories of Jack Strominger, Bill Lennarz, Phil Robbins, Mary Jane Osborn, and Bernard Horecker. The first lipid intermediate was detected in bacteria, and it was shown to contain eleven isoprene residues, one of which was linked to pyrophosphate, which, in turn, was linked to sugar residues. In a series of steps, this was converted to a lipopolysaccharide which was incorporated into the cell wall. Another project involved dolichol, a compound containing many isoprene units. With N. Behrens, Leloir found that, when phosphorylated, dolichol could interact with UDPG. He thought that the product was a glucosylated protein, but he was unable to prove it. Work in other laboratories showed that dolichol pyrophosphate-linked oligosaccharide could be transferred to newly synthesized polypeptides in the endoplasmic reticulum.

In 1970, Leloir received the wonderful news that he had been awarded the Nobel Prize in Chemistry for his work on carbohydrate synthesis, and he and his staff celebrated the occasion by drinking champagne from laboratory glassware. He was the third Argentinian to receive this award. He became an instant hero in Argentina, with a postage stamp issued in his honor. In his career, he received numerous other honors, including membership in the National Academy of Sciences of U.S.A., the American Academy of Arts and Sciences, the American Philosophical Society, the Royal Society of Great Britain, the French Academy of Sciences, the Academia Nacional de Medicina, and the Pontifical Academy of Sciences. He received numerous honorary degrees and awards from the United States, Argentina, Canada, and Mexico—in particular the Gairdner Award.

Leloir said he found himself drawn to the relatively more tractable problems posed by biochemistry. He has been given credit for conducting world-class research under less than optimal conditions and with limited funding. He often used homemade apparatus and encouraged his staff to make inventions. One example is when he constructed makeshift gutters out of waterproof cardboard to protect his laboratory library from a leaky roof.

Leloir had a reputation for being courteous and accessible. One can almost discern this from his courtly features. He was somewhat leery of the Nobel, telling *Newsweek* that his prize money would be spent on further research "if I'm ever allowed to work again in the peace and quiet that I'm used to." He was also known for his humility. Asked about the significance of his achievement, Leloir responded: "This is only one step in a much larger project. I discovered (no, not me: my team) the function of sugar nucleotides in cell metabolism. I want others to understand this, but it is not easy to explain: this is not a very noteworthy deed, and we hardly know even a little." In his acceptance speech in Stockholm, he adapted Winston Churchill's speech to the House of Commons in 1940, remarking "never have I received so much for so little." Leloir's humility was in striking contrast to the behavior of some Laureates. He attended his institute daily until his death on December 4, 1987.

References

Behrens, N. H., and L. F. Leloir. 1970. "Dolichol Monophosphate Glucose: An Intermediate in Glucose Transfer in Liver." *Proc. Natl. Acad. Sci. U.S.A.* 66: 153–59.

Braun-Menendez, E., J. C. Fasciolo, L. F. Leloir, and J. M. Muñoz. 1940. " The Substance Causing Renal Hypertension" *J. Physiol.* 98: 283–98.

Cabib, E., and L. F. Leloir. 1954. "Guanosine Diphosphate Mannose." *J. Biol. Chem.* 206: 779–90.

Cabib, E., L. F. Leloir, C. E. Cardini. 1953. "Uridine Diphosphate Acetylglucosamine." *J. Biol. Chem.* 203: 1055–70.

Caputto, R., L. F. Leloir, C. E. Cardini, and A. C. Paladini. 1950. "Isolation of the Coenzyme of the Galactose Phosphate—Glucose Phosphate Transformation." *J. Biol. Chem.* 184: 333–50.

Caputto, L. F. Leloir, C. E. Trucco, E. Cardini, and A. C. Paladini. 1948. "A Coenzyme for Phosphoglucomutase." *Arch. Biochem.* 18: 201–03.

Cardini, C. E., L. F. Leloir, and J. Chiriboga. 1955. "The Biosynthesis of Sucrose." *J. Biol. Chem.* 214: 149–55.

Espada, J. 1962. "Enzymatic Synthesis of Adenosine Diphosphate Glucose from Glucose-1-Phosphate and Adenosine Triphosphate." *J. Biol. Chem.* 237: 3577.

Fasciolo, J. C., Luis F. Leloir, J. M. Muñoz, and E. B. Menendez. 1940. "On the Specificity of Renin." *Science* 92: 554–55.

Leloir, L. F. 1983. "Far Away and Long Ago." *Ann. Rev. Biochem* 52: 1–15.

Leloir, L. F., and J. M. Muñoz. 1939. "Fatty Acid Oxidation in Liver".*Biochem. J.* 33: 734–46.

Leloir, L. F. 1971. "Two Decades of Research on the Biosynthesis of Ssaccharides." *Science* 172: 1299–1303.

Niemeyer, H. 1955. *Metabolismo de los Hidratos de Carbono*. Santiago: Universidad de Chile.

Recondo, E., and L. F. Leloir. 1961. "Adenosine Diphosphate Glucose and Starch Synthesis." *Biochem. Biophys. Res. Commun.* 6: 85–88.

7

Earl Sutherland

Master of Intuition

Education and Rediscovery of Glucagon

Earl Wilbur Sutherland was born on November 19, 1915, in Burlingame, Kansas, a small agricultural town. He was the fifth of six children. His father was a merchant for forty years in Burlingame where he ran a dry goods store, at which his wife and children helped out. He was from Wisconsin and went to Grinnell College for two years before farming in New Mexico and Oklahoma. Earl's mother was from Missouri where she attended a "ladies college" and trained as a nurse. Earl's younger brother became a professor of chemical engineering at the University of Missouri, and an older half-brother became a Unitarian minister. The family was prosperous until the Great Depression brought hard times. Earl's mother had the most influence during his early years. She recognized and encouraged his interest in science and brought him up to be very independent. That independent streak remained throughout his life. She taught him to swim and let him fish in a nearby river. He was soon out in the woods and fields hunting rabbits and squirrels. These early experiences imbued in him a lifelong love of the outdoors and fishing.

In high school he played football and basketball and excelled in tennis. In 1933, Sutherland entered Washburn College in Topeka, Kansas, and had to support himself working as a hospital orderly. Because of the austerity that he experienced, he decided to become a doctor rather than a scientist. On graduating with a BS, he married Mildred Rice of Topeka. He then entered Washington University Medical School, and his performance in the pharmacology course so impressed Carl Cori that he offered him a student assistantship. Cori commented on the midwestern qualities that Sutherland exemplified in his outlook, sense of humor, conviviality, honesty, and directness, and felt that Earl remained a midwesterner all his life. Cori noted Sutherland's aversion to big-city life and his tendency to hold back from dealing with people if he was suspicious of their motives. It was not because he lacked sophistication, but rather because he preferred the simplicity and informality of his upbringing. Cori noted that it was not apparent early on that Sutherland was quite above the ordinary. Cori felt that the Midwest represented a great reservoir of talent, which drained East and West because of a greater challenge to ambition and the bigger opportunities elsewhere.

The position that Cori gave Sutherland when he was a medical student in 1940–1942 allowed him to try his hand in research. He worked with Sidney Colowick to synthesize glycogen from glucose in several steps using a combination of enzymes. These included hexokinase, which formed glucose-6-phosphate, phosphoglucomutase that converted glucose-6-phosphate to glucose-1-phosphate, and phosphorylase that formed glycogen from glucose-1-phosphate. This was the first recapitulation of a cellular metabolic process using purified enzymes.

After he received his MD in 1942, Sutherland worked as an intern at Barnes Hospital in St. Louis for one year and continued his research with Cori. The clinicians complained of Sutherland's frequent absences, but Cori defended him. During 1943–1944, Sutherland served in the U.S. Army Medical Corps and was assigned as a battalion surgeon to General George Patton's Third Army in Germany. On his return to St. Louis, Carl Cori convinced Sutherland to go into medical research and not clinical medicine, and Sutherland said it was a decision he never regretted. He stayed in the Cori department from 1945 to 1953, rising from instructor to associate professor. He found the interactions with the gifted investigators there to be very stimulating. In his Nobel address, he stated that a necessary critical mass of young and talented investigators, with the opportunity for the free exchange of ideas, is an important ingredient in the making of scientific progress.

The first project that Sutherland embarked on when he returned to St. Louis to work with Carl Cori was to explore why certain insulin preparations caused a breakdown of liver glycogen yet lowered the blood glucose, and why some preparations even caused hyperglycemia. The hypothesis was that even highly purified amorphous or crystalline insulin contained a contaminating glycogenolytic factor. He found that he could destroy the hypoglycemic effect of "insulin" with alkali or cysteine without altering the glycogenolytic effect, indicating that different factors were involved. More convincing was the fact that an insulin preparation from a Danish manufacturer (Novo Laboratories) could reduce the blood sugar without causing glycogenolysis. He thus concluded that a separate factor was responsible for the glycogenolysis and hyperglycemia. This was called the hyperglycemic-glycogenolytic factor (H-G factor), and he reasoned that it had to be present in the pancreas, which was the source from which all insulin preparations were made. In collaboration with Christian de Duve, a fellow from Belgium (see chapter 13), Sutherland found that a factor identical with that contaminating insulin was present in extracts from the pancreas and also from the mucosa of the stomach. The distribution of the factor in the pancreas followed roughly the distribution of the islets. In an early demonstration of his intuitive powers, he speculated (correctly) that the α-cells of the pancreas were the source of the factor and that it may have a physiological role. In the latter case, he was careful to point out that there needed to be proof that the factor was secreted into the bloodstream and actually participated in the regulation of the blood sugar. The H-G factor was later renamed when it was found to be identical with glucagon, which had been discovered in 1923 (not long after the discovery of insulin) as a contaminant of insulin. When Sutherland developed an assay, the factor was shown to be present in peripheral and pancreatic blood.

In collaboration with Carl Cori and others, Sutherland proceeded to purify the H-G factor from insulin preparations and the gastric mucosa. He decided to first test if phosphorylase was the enzyme stimulated by the factor and also by epinephrine. They had reasoned earlier that phosphorylase was the rate-limiting enzyme for glycogen breakdown, but they showed further, by addition of glucose-6-phosphate and glucose-1-phosphate to liver slices, that phosphoglucomutase and glucose-6-phosphatase were not affected by the glycogenolytic agents. Furthermore, using ^{32}P-inorganic phosphate to label glucose-1-phosphate and glucose-6-phosphate in the slices, they showed that both were increased by glucagon and epinephrine, consistent with an action on phosphorylase. To clinch their hypothesis, they measured phosphorylase in homogenates prepared from liver slices and found that pretreatment of the slices with the two agents increased its activity. They reasoned, correctly, that liver contained an enzyme system that keeps a balance between an active and inactive form of phosphorylase, permitting rapid change in either direction, and that this system is influenced by glycogenolytic agents. This conjecture was an example of Sutherland's remarkable intuition.

Sutherland's focus was not entirely on the mechanisms controlling phosphorylase. He also worked with Theo Posternak, Mildred Cohn, and Carl Cori to elucidate the reaction mechanisms for two glycolytic enzymes, phosphoglucomutase and phosphoglyceromutase. Their interesting findings are described in the next chapter.

Discovery of Cyclic AMP

In 1953, Sutherland moved to Cleveland to become the chair of pharmacology at Western Reserve University (now Case Western University), and there was a significant gap in his research before he engaged in a detailed study of the inactivation and activation of liver phosphorylase and characterized the enzymes involved and how they were affected by glucagon and epinephrine. Despite Carl Cori's misgivings, Sutherland decided to develop an in vitro system to explore the mechanism of action of the hyperglycemic factors using broken cell preparations (liver homogenates) and purified enzymes. Cori felt, as did many others, that the mechanism of action of any hormone would be lost once its target cells were disrupted. But Sutherland defied conventional wisdom and went ahead anyway. This manifestation of his independent spirit would result in a discovery of great biological and medical significance.

In Cleveland, Sutherland's focus was on the enzymes that inactivated and activated phosphorylase. The inactivating enzyme was purified from liver and was found to release phosphate from phosphorylase—that is, it had phosphatase activity but not toward simple phosphate esters. Importantly, the release of phosphate from the enzyme correlated with loss of its activity. Through the use of liver slices incubated with ^{32}P-inorganic phosphate to label the ATP, the label was found to be incorporated into phosphorylase in a form that could not be released by acid treatment; in other words, it was covalently bound. Furthermore, this activity was observed to be consistently stimulated by epinephrine and glucagon. These findings indicated that the activity of phosphorylase was controlled by a phosphorylation/dephosphorylation mechanism. Sutherland also showed that a similar system operated in heart muscle.

A critical phase in the research came when Sutherland and his associates studied the activation of phosphorylase by epinephrine and glucagon in cell-free liver homogenates. They purified the activating enzyme and found, as confirmed by previous findings, that it was a kinase. Consistent with this, its activity was shown to be dependent on ATP and Mg^{2+} ions. During the progress of the work, Edwin Krebs and Edmond Fischer at the University of Washington in Seattle had discovered that the conversion of phosphorylase b to phosphorylase a in muscle also involved a phosphorylation reaction catalyzed by a protein kinase (phosphorylase kinase). This is described in detail in the chapter 11 on Edwin Krebs.

Sutherland, W. D. Wosilait, and Jacques Berthet, a fellow from Louvain, Belgium, then initiated a series of studies in which hormones were added directly to liver homogenates in the presence of ATP and Mg^{2+} ions to test for activation of phosphorylase. It is of historical interest to note that Sutherland had tried for years without success to see hormone effects in liver homogenates. This was a gamble because in all previous experiments, the hormones had been added to intact liver slices. The key difference was the addition of ATP and Mg^{2+} ions. Sutherland finally did observe phosphorylase activation in crude liver homogenates in the presence of these additions. Since phosphorylase was present in the clear supernatant obtained by centrifuging the homogenates, they tested this fraction for hormone effects. Alas, none could be seen. In an example of Sutherland's intuitive thinking, the researchers carried out a simple additional experiment. They added back the sedimented particulate fraction to the supernatant and found that hormone activation of phosphorylase reappeared. They then incubated the particulate fraction with the hormones and found that a factor was produced that could activate phosphorylase when added to the supernatant fraction. Thus, by proceeding along a simple logical train of thought, they determined that the hormone response could be broken into two consecutive reactions—first, production of the factor in the particulate fraction in the presence of the hormones and, second, the activation of phosphorylase by the factor in the supernatant fraction.

The focus of the research then turned to identifying the factor. This was found to be dialyzable and heat stable, but identifying it proved very difficult because of its low concentration in tissues and the presence of an enzyme that destroyed it. However, Sutherland, in association with Theodore (Ted) Rall, proceeded to purify it from liver particulate fractions using ion exchange chromatography and phosphorylase activation as an assay. Chemical and UV analysis showed it to contain adenine, ribose, and phosphate in a 1:1:1 ratio, but it differed from all known adenine nucleotides. Its identification was made as a result of a remarkable case of serendipity, described as follows by Maxine Singer. Sutherland wrote to Leon Heppel, an expert on nucleotides at NIH, for help in identifying the factor. But Heppel was not a good correspondent who let his letters accumulate on his desk. Periodically, one of his staff was assigned to clear off his desk—a dreaded task since unopened and opened mail were mixed together and layered with brown paper! On one famous occasion, Heppel rushed down the corridor after opening Sutherland's letter and excavating through layers of mail until he found one from David Lipkin's group at Washington University. Lipkin had written to Heppel describing a new nucleotide produced by treating ATP with barium hydroxide. Heppel realized that both investigators were probably dealing with the same compound, so he sent letters to both of them. Sure enough, the compounds were found to be the same cyclic mononucleotide, adenosine 3', 5'-monophosphate (now usually called cyclic AMP). Lipkin later provided a complete proof of its structure.

The production of large amounts of cyclic AMP by the barium hydroxide method was a great boon for Sutherland's later studies. With Rall, Sutherland found that tissues contained an enzyme that inactivated the nucleotide, and he surmised, correctly, that it was a phosphodiesterase. This was in line with Sutherland's logic that any mechanism involving hormone action must have a system for terminating the excitatory activity. Sutherland also proceeded to identify the enzyme that formed cyclic AMP. For this, a crude assay system for cyclic AMP was used, namely the activation of phosphorylase in liver homogenates, with changes in glycogen measured with iodine using a simple Klett-Summerson colorimeter. This assay system often failed, leading to much frustration, but was used for many years until Alfred Gilman devised a more accurate and sensitive system based on the binding of cyclic AMP to cyclic AMP-dependent protein kinase. Gilman was an MD, and PhD student in Sutherland's laboratory who won the Nobel Prize in 1994 for his discovery of guanine nucleotide binding regulatory proteins. These are now abbreviated as G proteins and are essential components in signal transduction for many hormones, that is, transmission of the signal from their surface receptors to intracellular enzyme systems.

Studies on Adenylyl Cyclase

The enzyme that converts ATP to cyclic AMP was initially named adenyl cyclase. Waldo Cohn, an authority on nomenclature based at the Oak Ridge National

Laboratory, sarcastically criticized this name as chemically incorrect. Sutherland had named the enzyme without much thought and willingly agreed to rename it adenylyl cyclase, although he found it amusing that someone could get worked up about such an issue. The enzyme was characterized and found to be widely distributed in mammals and birds and was even found in lower orders such as flies, liver flukes, and worms. Its distribution in mammalian tissues was also widespread, with the highest activity in brain cerebellum and cortex. These findings indicated that the regulation of cellular processes by cyclic AMP could be very widespread. In association with Rall, Sutherland purified adenylyl cyclase from a particulate fraction of cerebral cortex. The nature of the particulate material remained unknown, but it was suggested that it could be derived from cell membranes or nuclei.

The mechanism of the cyclase reaction was explored. As expected, pyrophosphate was the other product of the reaction, and using ATP labeled in different phosphates with ^{32}P, it was shown that the α-phosphate was incorporated into cyclic AMP, whereas the β- and γ-phosphates went to pyrophosphate. In association with Ferid Murad, an MD, PhD student, Rall and Sutherland looked at the effects of various catecholamines (the D-isomers of epinephrine and norepinephrine and a pharmaceutical analog isoproterenol) on the cyclase activity of preparations from heart and liver. The results showed a clear potency series of isoproterenol > epinephrine > norepinephrine, indicating that the catecholamines interacted with a receptor of the β-adrenergic type.

Sutherland started to look at the effects of epinephrine on adenylyl cyclase preparations from various tissues. Effects were seen in most parts of the brain and lung, skeletal and smooth muscle, arteries, spleen, and fat tissue. Surprisingly, avian erythrocytes were tested and showed a large response. This effect, which was not seen in mammalian red cells, proved useful in defining the cellular source of the enzyme. In association with postdoctoral fellows and external collaborators, Sutherland initiated a large series of experiments involving other hormones and their target tissues. These indicated the widespread involvement of cyclic AMP as a mediator of hormone actions. In addition to the effects of glucagon and epinephrine noted above, other hormones were found to increase the tissue level of cyclic AMP, namely vasopressin on the bladder; adrenocorticotrophic hormone (ACTH) on the adrenal cortex; luteinizing hormone (LH) on the corpus luteum; and epinephrine on the heart, adipose tissue, and intestinal smooth muscle. Furthermore, it was found that in a single tissue (adipose tissue), many agents increased cyclic AMP, including epinephrine, norepinephrine, ACTH, LH, glucagon, thyroid-stimulating hormone, and prostaglandin E_1. All of these agents caused lipolysis (breakdown of fat in adipose tissue to release fatty acids), thus linking cyclic AMP to an important physiological response. These findings suggested that diverse hormones could act on the same cells to raise cyclic AMP through a common pathway. Sutherland even found cyclic AMP in *Escherichia coli*, where its production was increased by glucose deprivation.

Move to Vanderbilt University

During the first phase of this work, Sutherland was having personal problems partly due to a divorce and partly due to his situation at Western Reserve. The medical school was undergoing a major change in its curriculum, and Sutherland, who never enjoyed lecturing and administration, wanted to get into a situation where he could devote his energy almost entirely to research. Vanderbilt University in Nashville, Tennessee, where some of his former colleagues (Sidney Colowick, Charles and Janey Park, Victor Najjar) were located, provided him with such an environment. He was appointed to the Physiology Department and awarded a Career Investigatorship of the American Heart Association. Shortly before moving to Vanderbilt, Sutherland married Claudia Sebeste, a lady of Hungarian extraction with great charm. He shared with her his passion for fishing, which the lakes and streams around Nashville provided in abundance. He claimed that some of his best scientific thoughts came while on the water.

Some of his fellows (R. W.(Bill) Butcher and G. A.(Al) Robison) went with him to Vanderbilt, and they played an integral role in the studies of hormone effects in various tissues. Measurement of cyclic AMP still involved the tedious method of measuring the activation of phosphorylase in liver homogenates. Because of the large amount of tissue needed, the livers were taken from dogs, and Sutherland's long-term technician, James Davis, went with him from Cleveland to perform the grisly task of sacrificing the animals and homogenizing their livers. This was carried out in the basement of the medical school, well away from the prying eyes of animal rights groups. This was before Gilman's much improved assay was available.

The fact that avian erythrocytes showed a cyclic AMP response to epinephrine provided the opportunity to find the origin of the particulate fraction that generated the nucleotide. This is because plasma membranes can be prepared from these cells by the simple procedures of fragmentation and centrifugation. In studies carried out by Ivar Øye, a fellow from Oslo, epinephrine-sensitive adenylyl cyclase was found to be associated with the membranes. Since intact erythrocytes did not respond, it was deduced that the responsive enzyme system was on the inner surface of the membranes.

During Sutherland's classic studies with Rall on the formation of cyclic AMP, he noted that there was an activity that destroyed the nucleotide and found that it was inhibited by caffeine. He studied this activity and found that it was due to a phosphodiesterase that yielded 5'-AMP. With Bill Butcher, who was a graduate student at that time, he set out to purify and characterize the enzyme. It was found to be specific for cyclic 3′, 5′-nucleotides and to be present in all tissues expressing adenylyl cyclase activity. No other mechanism for the destruction of cyclic AMP was found. The inhibitory effect of caffeine on the phosphodiesterase represented the first demonstration of how caffeine and related methylxanthines exerted their effects.

Sutherland surmised that other cyclic nucleotides probably existed. This proved to be the case when another investigator, T. D. Price found cyclic GMP in urine. In a new project initiated at Vanderbilt with Joel Hardman, a postdoctoral fellow, Sutherland devised a method for measuring cyclic GMP and other cyclic nucleotides (Figure 12). This demonstrated the presence of both cyclic AMP and cyclic GMP in the urine of rats and showed that the level of cyclic GMP was altered by the presence or absence of pituitary and certain other hormones.

With Hardman, Sutherland proceeded to characterize guanylyl cyclase, the enzyme that catalyzed the formation of cyclic GMP from GTP. It soon became clear that it was different from adenylyl cyclase. For example, it was largely soluble, and its distribution in several tissues was quite different. It was not activated by glucagon, fluoride, and epinephrine under conditions where adenylyl cyclase was activated, and its activity was much greater with Mn^{2+} than Mg^{2+}. In a later study in collaboration with Günter and Karin Schultz from Heidelberg, Hardman and Sutherland showed that carbachol, a parasympathetic agonist, increased cyclic GMP in the smooth muscle of the ductus deferens. Subsequent work in other laboratories showed that the enzyme was stimulated by nitrites and other agents that produced nitric oxide (NO), and the effect of carbachol was attributed to its stimulation of the production of NO. The surprising recognition that a toxic gas such as NO could act as a hormone resulted in the award of the Nobel Prize in 1998 to Ferid Murad, Robert Furchgott, and Louis Ignarro.

In 1966, Sutherland and Robison formulated their famous second messenger hypothesis in an article in *Pharmacological Reviews*. This hypothesis proposed that a hormone (the first messenger) interacted with the outside of the cell to activate adenylyl cyclase in the membrane to increase the level of cyclic AMP (the second messenger) inside the cell. The scheme left open the possibility of other second messengers generated by other hormones. Sutherland developed certain criteria to decide whether a hormone acted this way. These were similar to the postulates to define the bacterial basis of diseases developed by the German physician Robert Koch, who isolated the bacteria responsible for anthrax, tuberculosis, and cholera. In his adaptation of Koch's postulates, Sutherland stated that first, stimulation of adenylyl cyclase by a hormone should be demonstrable in intact tissues as well as broken cells. Second, the changes in the cyclic AMP concentration in tissues should reflect the dose response curve for the action of the hormone and show a proper time relationship to its action. Third, agents that inhibit cyclic AMP phosphodiesterase, such as caffeine and theophylline, should potentiate the hormone action. Fourth, it should be possible to mimic the hormone effect by adding cyclic AMP or its lipophilic derivatives to tissue preparations, although in practice this was often difficult.

Studies on Phosphodiesterase and the Action of Insulin

Sutherland returned his attention to phosphodiesterase. With Hardman and a graduate student named Joe Beavo, Sutherland made a detailed kinetic study of the enzyme with cyclic AMP and cyclic GMP as substrates. Their data suggested, correctly, the existence of more than one phosphodiesterase in a tissue. In most cases, the two nucleotides interfered with the hydrolysis of each other in a predictable manner, but in one case, cyclic GMP stimulated rather than inhibited the hydrolysis of cyclic AMP. In his later career, Beavo would identify and characterize a large number of cyclic nucleotide phosphodiesterases. Sutherland also wanted to characterize adenylyl cyclase, but had been hampered by its insolubility and instability. Together with a postdoctoral fellow, Roger Johnson, Sutherland found that adenylyl cyclase could be dispersed by the detergent Lubrol-PX with retention of activity. This enabled studies of the effects of fluoride and divalent metal ions.

Sutherland was interested in the mechanism of action of insulin, and this was stimulated when he moved to Vanderbilt, where Charles Park and associates had done pioneering work on the mechanism by which insulin stimulated glucose uptake into cells. In association with Butcher, he showed that insulin could reduce the level of cyclic AMP in adipose tissue when this was raised by epinephrine, ACTH or glucagon, but had little effect in their absence. The ability of insulin to decrease cyclic AMP under certain circumstances was reinforced by the work of L. S. (Jim) Jefferson, a graduate student. He found that treatment of rats with anti-insulin serum or with alloxan to induce diabetes resulted in an increase in cyclic AMP in the liver, which was reversed by treatment with insulin. Later experiments conducted by myself in association with Robison, Park, and Stephen Lewis, a very energetic medical doctor, showed that insulin inhibited the metabolic actions of glucagon on the liver by lowering the level of cyclic AMP. Sutherland believed that the second messenger for insulin was cyclic GMP, but in this instance his intuition was wrong. Many experiments failed to show that insulin raised cyclic GMP. The mechanism by which insulin opposed the actions of hormones that raise cyclic AMP became clear when a postdoctoral fellow, J. G. T. (Sam) Sneyd, who previously worked with Park and Butcher, showed that insulin stimulated cyclic AMP phosphodiesterase in fat cell extracts.

Studies of Cyclic Nucleotides in Humans

Sutherland never forgot his roots in medicine and, together with Grant Liddle, the chairman of medicine at Vanderbilt, conducted an extensive series of studies of cyclic nucleotides in man. These involved Hardman and several research fellows, together with Arthur Broadus an MD, PhD student. Hardman had developed a new and much improved assay for cyclic AMP and cyclic GMP in plasma and urine, which was used in the studies and which also involved administering radioactive cyclic AMP and cyclic GMP by intravenous catheters into human volunteers. Both cyclic nucleotides were found to be cleared in the kidneys by glomerular filtration. The kidneys were also observed to produce endogenous cyclic AMP, but not cyclic GMP. The plasma levels of both nucleotides were noted to be in a dynamic state, with clearance being due mainly to extrarenal processes involving phosphodiesterase. It is interesting to note that many volunteers for these studies were students or inmates at the Tennessee State Penitentiary. The latter were eager to relieve the tedium of prison life and, in many cases, were found to be of high intelligence. Such studies would not be permitted today.

When glucagon was administered, plasma and urinary cyclic AMP were increased, but cyclic GMP was unaffected. Further experiments demonstrated that the liver was the source of the additional cyclic AMP. Injection of parathyroid hormone increased both plasma and urinary cyclic AMP. Most of the urinary cyclic AMP arose by glomerular filtration, but a significant fraction was produced by the kidneys. Epinephrine, norepinephrine, and isoproterenol increased plasma and urinary cyclic AMP but did not affect plasma cyclic GMP. Using specific blockers of α-adrenergic or β-adrenergic receptors, further experiments showed that the increases in cyclic AMP were mediated by β-adrenergic receptors in accord with findings in isolated tissues. Interestingly, when the β-adrenergic effects of the catecholamines were blocked so that only α-adrenergic effects were seen, there was a rise in plasma cyclic GMP, but the mechanisms underlying this were unclear. This was because in such studies performed in intact humans, the effects of α-adrenergic stimulation could have been due to secondary changes, for example, stimulation of the parasympathetic system.

Sadly, during his time at Vanderbilt, Sutherland's health had been declining, and it became difficult for him to present the many lectures that he was invited to give. Often, his postdoctoral students had to cover for him and give his lectures, and they became very adept at this. His long-anticipated Nobel Prize for the discovery of cyclic AMP was awarded in 1971. When a crew from Swedish television arrived in Nashville a few days prior to the announcement, we knew his day had finally come. Like Herman Kalckar and Fritz Lipmann, Sutherland was not a good speaker and really disliked having to give talks. In 1973, Sutherland made a surprising move to the University of Miami, but his failing health meant that he accomplished little scientifically. He died of esophageal bleeding nine months later on March 9, 1974, at the early age of 59.

Sutherland's Awards and Reminiscence by Joel Hardman

Sutherland received many stellar awards besides the Nobel Prize. These include the Lasker Award, the National Medal of Science, the Gairdner Award, the Banting

Medal, the Dickson Prize, and the Sollman Award. He was elected to the National Academy of Sciences and to the American Academy of Arts and Sciences. He was also a Fellow of the American Association for the Advancement of Science and was given a Research Achievement Award from the American Heart Association. The success of his predoctoral and postdoctoral students was impressive; two of them, Alfred Gilman and Ferid Murad, were awarded Nobel Prizes. Other students of Sutherland's were also mentored as young faculty by Joseph Larner, and their careers are described in a later chapter devoted to Larner. Like Gilman and Murad, other members of Sutherland's laboratory moved on to occupy chairs or high administrative positions in medical schools. Thus Butcher became chair of biochemistry at the University of Massachusetts Medical School and later moved to the University of Texas, Houston, where he became chair of biochemistry and dean of graduate studies. Robison became chair of physiology at the University of Texas, Houston; Hardman became chair of pharmacology at Vanderbilt University and later associate vice chancellor for Research; and Günter Schultz chaired the Institute of Pharmacology at the Free University of Berlin.

I am indebted to Joel Hardman for the following personal reminiscences about Sutherland, to which I have added the thoughts of Bill Butcher and Al Robison, two of his early associates, and a few of my own. Joel joined Sutherland's laboratory at Vanderbilt in 1964 and served as his right-hand man until Sutherland left for Miami in 1973. He notes that Sutherland was a complex man, even by the high standards of highly creative, intelligent, and successful people. He cherished independence in his thoughts and actions and was too much an independent thinker to form a fast attachment to a political party or dogmatic religion. He could be a hawk or a dove, but he was not predictably either one. He was generous to and supportive of his friends and the people who worked in his laboratory. Social events at the Sutherland home were frequent and memorable because of the sincere and warm hospitality shown by Earl and Claudia. Earl often held the rapt attention of dinner guests with stories of his wartime experiences in Patton's Third Army.

Sutherland loved science, football, gardening, fishing, conversation, whiskey, and his Labrador retrievers, but not necessarily in that order. The dogs had flunked obedience school numerous times and had the run of the house whether dinner guests were present or not. Commonly, guests would leave with their clothes coated with saliva as the dogs dropped balls or sticks in their laps for them to throw. It never occurred to Earl that his guests might not like his dogs as much as he did. On those rare occasions when the dogs were put outside, they would hurl themselves against the glass doors, and the guests would imagine being showered with glass shards.

Hardman describes Sutherland's scientific instincts as remarkable. It almost seemed as if he could peer into a living cell and see what occurred and why. He had an uncanny knack for asking the right question and getting the answer by the most direct and efficient route. He would spend as long as it took to convince himself that he was asking the right question and then to design the experiments that would yield results that would give a yes or no answer. He could spot trends in the data of associates that were not apparent to them, and he often left them embarrassed when he pointed out that some of the present data were inconsistent with data shown to him weeks or months earlier. He stored information with amazing efficiency, as though he had an intuitive filter that only stored information that was truly significant. He never drew conclusions prematurely and could find alternative interpretations for results that seemed straightforward. Hypotheses seemed to him to be either worthy or unworthy of testing. He seemed to have a sixth sense about how biological systems ought to work, but he kept a strongly dispassionate objectivity when he tested a favorite hypothesis. A precept by which he lived, and what he imparted to his colleagues, was that you never fell in love with your hypothesis. He seldom read scientific papers unless recommended by a colleague. He used scientific meetings and conversations as his major sources of information, and he commonly knew about most of the developments in his field well before they appeared in print. This was because other workers would often call him to seek his opinion and advice.

Sutherland disliked classroom teaching and did not give good lectures or seminars. Charles Park, who was chair of the Physiology Department, recognized Sutherland's limitations as a lecturer in the medical course, but said: "I feel that the students should at least see him!" In contrast to his formal presentations, he was a superb teacher in the laboratory and in small group discussions. He showed how research should be carried out intellectually, with careful planning of protocols, rigorous challenging of data interpretation, and unfailing attention to the question being asked. During discussions, Sutherland always encouraged his young associates with respect to the work accomplished and the quality of the data. However, he was much more selective in his praise of ideas and concepts. If he thought an idea was half-baked, he would patiently point out its shortcomings. If he thought an idea or experimental design was simply terrible, he usually would not respond verbally, but would lean back in his chair with a faraway glazed-over look in his eyes. This would convey as much unequivocal disdain as the most scathing verbal critique. He was a strong believer in the value of open scientific conversation. There were never bits of secret information in his group. In fact, secrets were anathema to him since they violated his concept of what science should be. He seemed to be motivated primarily by a strong desire to understand things rather than the desire to be the first to discover something new.

In scientific conversations and lectures, Sutherland's thoughts occasionally outran his words. He would assume at times that those to whom he was talking knew about past events in the laboratory, whether they did or not. In this way he resembled Herman Kalckar whose communication characteristics have been described earlier. I personally experienced this in conversations with Sutherland when he assumed I knew about previous experiments. He was prone to using unusual abbreviations and would switch topics, sometimes in mid-sentence. Often I would emerge from the meetings mentally exhausted and doubting my intelligence. Sutherland was unquestionably a brilliant man, perhaps a genius, but by no means could he be called an intellectual. He was too pragmatic to be interested in any form of the arts or great literature. As Carl Cori stated at the beginning of this chapter, despite all of his awards and accolades, Sutherland retained the character and values of a midwesterner.

References

Ball, J. H., N. I. Kaminsky, J. G. Hardman, A. E. Broadus, E. W. Sutherland, and G. W. Liddle. 1972. "Effects of Catecholamines and Adrenergic Blocking Agents on Plasma and Urinary Cyclic Nucleotides in Man." *J. Clin. Invest.* 51: 2124–29.

Beavo, J. A., J. G. Hardman, and E. W. Sutherland. 1970. "Hydrolysis of Cyclic Guanosine and Adenosine 3',5'-Monophosphates by Rat and Bovine Tissues." *J. Biol. Chem.* 245: 5649–55.

Broadus, A. E., N. I. Kaminsky, J. G. Hardman, E. W. Sutherland, and G. W. Liddle. 1970. "Kinetic Parameters and Renal Clearances of Plasma Adenosine 3',5'-Monophosphate and Guanosine 3',5'-Monophosphate in Man." *J. Clin. Invest.* 49: 2222–36.

Broadus, A. E., N. I. Kaminsky, R. C. Northcutt, J. G. Hardman, E. W. Sutherland, and G. W. Liddle. 1970. "Effects of Glucagon on Adenosine 3',5'-Monophosphate and Guanosine 3',5'-Monophosphate in Human Plasma and Urine." *J. Clin. Invest.* 49: 2237–45.

Butcher, R. W., C. E. Baird, and E. W. Sutherland. 1968. "Effects of Lipolytic and Antilipolytic Substances on Adenosine 3',5'-Monophosphate Levels in Isolated Fat Cells." *J. Biol. Chem.* 243: 1705–12.

Butcher, R. W., J. G. Sneyd, C. R. Park, and E. W. Sutherland. 1966. "Effect of Insulin on Adenosine 3',5'-Monophosphate in the Rat Epidimal Fat Pad." *J. Biol. Chem.* 241: 1651–53.

Butcher, R. W., and E. W. Sutherland. 1962. "Adenosine 3',5'-Phosphate in Biological Materials. I. Purification and Properties of Cyclic 3',5'-Nucleotide Phosphodiesterase and Use of the Enzyme to Characterize Adenosine 3',5'-Phosphate in Human Urine." *J. Biol. Chem.* 237: 1244–50.

Colowick, S. P., and E. W. Sutherland. 1942. "Polysaccharide Synthesis from Glucose by Means of Purified Enzymes." *J. Biol. Chem.* 144: 423–37.

Cori, C. F. 1978. "Earl W. Sutherland 1915–1974." *Biogr. Mem. Natl. Acad. Sci. U.S.A.* 49: 318–50.

Davoren, P. R., and E. W. Sutherland. 1963. "The Effect of L-Epinephrine and Other Agents on the Synthesis and Release of Adenosine 3',5'-Phosphate by Whole Pigeon Erythrocytes." *J. Biol. Chem.* 238: 3009–15.

Hardman, J. G., J. W. Davis, and E. W. Sutherland. 1969. "Effects of Some Hormonal and Other factors on the Excretion of Guanosine 3',5'-Monophosphate and Adenosine 3',5'-Monophosphate in Rat Urine." *J. Biol. Chem.* 244: 6354–62.

Hardman, J. G., and E. W. Sutherland. 1969. "Guanyl Cyclase, an Enzyme Catalyzing the Formation of Guanosine 3',5'-Monophosphate from Guanosine Triphosphate." *J. Biol. Chem.* 244: 6363–70.

Jefferson, L. S., J. H. Exton, R. W. Butcher, E. W. Sutherland, and C. R. Park. 1968. "Role of Adenosine 3',5'-Monophosphate in the Effects of Insulin and Anti-Insulin Serum on Liver Metabolism." *J. Biol. Chem.* 243: 1031–38.

Johnson, R. A., and E. W. Sutherland. 1973. "Detergent-Dispersed Adenylate Cyclase from Rat Brain. Effects of Fluoride, Cations, and Chelators." *J. Biol. Chem.* 248: 5114–21.

Kaminsky, N. I., A. E. Broadus, J. G. Hardman, D. J. Jones Jr., J. H. Ball, E. W. Sutherland, and G. W. Liddle. 1970. "Effects of Parathyroid Hormone on Plasma and Urinary Adenosine 3',5'-Monophosphate in Man." *J. Clin. Invest.* 49: 2387–95.

Klainer, L. M., Y.-M. Chi, S. L. Freidberg, T. W. Rall, and E. W. Sutherland. 1962. "Adenyl Cyclase. IV. The Effects of Neurohormones on the Formation of Adenosine 3',5'-Monophosphate by Preparations from Brain and Other Tissues." *J. Biol. Chem.* 37, 1239–43.

Loten, E. G., and J. G. T. Sneyd. 1970. "Effect of Insulin on Adipose Tissue Adenosine 3',5' Cyclic Monophosphate Phosphodiesterase." *Biochem. J.* 120: 187–93.

Murad, F., Y.-M. Chi, T. W. Rall, and E. W. Sutherland. 1962. "Adenyl Cyclase. III. Effects of Catecholamines and Choline Esters on the Formation of Adenosine 3',5'-Phosphate in Preparations from Cardiac Muscle and Liver." *J. Biol. Chem.* 237: 1233–38.

Price, T. D., D. F. Ashman, and M. M. Melicow. 1967. "The Effect of Epinephrine and Other Agents on Adenyl Cyclase in the Cell Membrane of Avian Erythrocytes." *Biochim. Biophys. Acta* 138: 452–65.

Rall, T. W., and E. W. Sutherland. 1958. "Formation of a Cyclic Adenine Ribonucleotide by Tissue Particles." *J. Biol. Chem.* 232: 1065–76.

Rall, T. W., and E. W. Sutherland. 1962. "Adenyl Cyclase. II. The Enzymatically Catalyzed Formation of Adenosine 3',5' Phosphate and Inorganic Pyrophosphate from Adenosine Triphosphate." *J. Biol. Chem.* 237: 1228–32.

Rall, T. W., E. W. Sutherland, and J. Berthet. 1957. "The Relationship of Epinephrine and Glucagon to Liver Phosphorylase. IV. Effect of Epinephrine and Glucagon on the Reactivation of Phosphorylase in Liver Homogenates." *J. Biol. Chem.* 224: 463–75.

Robison, G. A., R. W. Butcher, and E. W. Sutherland. 1968. "Cyclic AMP." *Ann. Rev. Biochem.* 37: 149–74.

Singer, M. 2003. "Leon Heppel and the Early Days of RNA Biochemistry." *J. Biol. Chem.* 278: 47351–56.

Sutherland, E. W., and C. F. Cori. 1948. "Influence of Insulin Preparations on Glycogenolysis in Liver Slices." *J. Biol. Chem.* 172: 737–50.

Sutherland, E. W., and C. F. Cori. 1951. "Effect of Hyperglycemic-Glycogenolytic Factor and Epinephrine on Liver Phosphorylase." *J. Biol. Chem.* 188: 531–43.

Sutherland, E. W., and C. de Duve. 1948. "Origin Distribution of the Hyperglycemic-Glycogenolytic Factor of the Pancreas." *J. Biol. Chem.* 175: 663–74.

Sutherland, E. W., C. F. Cori, R. Haynes, and N. S. Olsen. "Purification of the Hyperglycemic-Glycogenolytic Factor from Insulin and Gastric Mucosa." 1949. *J. Biol. Chem.* 180: 825–37.

Sutherland, E. W., I. Øye, and R. W. Butcher. 1965. "The Action of Epinephrine and the Role of the Adenyl Cyclase System in Hormone Action." *Recent Prog. Horm. Res.* 21: 623–46.

Sutherland, E. W., and T. W. Rall. 1958. "Fractionation and Characterization of a Cyclic Adenine Ribonucleotide Formed by Tissue Particles." *J. Biol. Chem.* 232: 1077–91.

Sutherland, E. W., T. W. Rall, and T. Menon. 1962. Adenyl Cyclase. I. Distribution, Preparation and Properties. *J. Biol. Chem.* 237: 1220–27.

Sutherland, E. W., and G. A. Robison. 1966. "The Role of Cyclic-3',5'-AMP in Responses to Catecholamines and Other Hormones." *Pharmacol. Rev.* 18: 145–61.

Sutherland, E. W., and W. D. Wosilait. 1955. "Inactivation and Activation of Liver Phosphorylase." *Nature* 175: 169–70.

Sutherland, E. W., and W. D. Wosilait. 1956. "The Relationship of Epinephrine and Glucagon to Liver Phosphorylase. II. Enzymatic Inactivation of Liver Phosphorylase." *J. Biol. Chem.* 218: 469–81.

Sutherland, E. W., and W. D. Wosilait. 1956. "The Relationship of Epinephrine and Glucagon to Liver Phosphorylase. III. Reactivation of Liver Phosphorylase in Slices and Extracts." *J. Biol. Chem.* 218: 483–95.

8

Coris' Move to the Department of Biological Chemistry

Award of Nobel Prizes and Career of Tom Cori

Life in the Cori Laboratory

The next phase of the Coris' life involved Carl's appointment as chairman of the Department of Biological Chemistry at Washington University in 1945. This was associated with the provision of more space, which allowed more young investigators to flock to their laboratory from parts of the United States and all over the world. The careers of many of these investigators, some of whom won Nobel Prizes, are described in other chapters in this book. The memoirs of two of these, Mildred Cohn and Arthur Kornberg, describe what it was like to work in the Cori department. For example, Kornberg describes the very different personalities of the Coris, with Carl being calm and analytical and Gerty being emotional and intuitive. Despite these differences, they worked as an effective team in which their differences complemented each other. In the later stages of their careers in St. Louis, Carl did not have his own laboratory and so conducted his experiments in Gerty's laboratory. Gerty remained active in the laboratory until almost her death, although Carl would be actively involved in the design and interpretation of her experiments. With her exuberant personality, she would rush into Carl's office in great excitement with the latest results. Both of them inspired others with their work ethic, optimistic outlook, and broad knowledge of biological science. Mildred Cohn also described how it was to be in a department where both Carl and Gerty were consistently supportive, respectful, encouraging, and appreciative. Like other people in the Cori department, she was impressed by the breadth and depth of Carl's knowledge, his fluency in several languages, and his remarkable memory. She noted that despite these formidable attributes, he could be lighthearted, and the daily meetings were often livened by his wit. She described him as "one of those rare individuals to whom all products of the human intellect are accessible, and that he was equally at home discussing archeology, music or botany.

He was intellectually and personally so compelling that even occasional contact with him left a tremendous impression." She goes on to say that "his direct, unornamented approach to the pursuit of his scientific objectives could make him seem aloof, even austere, but he was never solemn, and his high spirits often gave rise to a wonderful gaiety."

Sir Philip Randle, the late chair of clinical biochemistry at the University of Oxford, described the atmosphere of collaboration and mutual support in the original laboratory at Washington University and noted that it continued when the Coris moved to the larger laboratory in the Department of Biological Chemistry. At that time, all the departments of the university had their own libraries. In the case of the Coris, the library was right next to the laboratory, which was next to Carl's office. This indicated the importance they placed on keeping up with the literature. Seminars were held in the library—usually over lunch. Visiting scientists were invited to present their findings or comment on the work of departmental members. Seminar speakers used either a blackboard or glass slides. Attendance by all members of the department was required, and apologies were expected if absence was unavoidable. On these occasions, Carl's formidable intellect and memory would be on display. He would try to be scrupulously fair in criticizing the seminars, but Gerty would just indicate her approval or disapproval by the appropriate facial expression. All this tended to be intimidating, especially to the junior members. An important feature of the department was that, prior to submission, all papers were read and critiqued by all the members.

Randle noted that Carl remained in essence a European with wide cultural interests and did not appreciate it if any students had a narrow focus on their projects. He was a good speaker, but tended to be nervous at first. His talks reflected his clear and precise thinking. But he could be sensitive to criticism of his findings or ideas. Mildred Cohn noted that Carl's writing style was economical, and the logic and clarity of his thinking was formidable. The principle on which he ran his department was to have members of his group work on different aspects of the same project. Thus they would bring different approaches and ideas to the research and would benefit mutually from their interactions.

Career of Tom Cori

The Coris lived with their son Tom, who was born in 1936, in a pleasant modern-style house where they enjoyed gardening, a lifelong hobby. Gerty took care of the flowers, while Carl looked after the vegetables. Their son Tom was born in one of the hottest summers on record in St. Louis, and, despite the lack of air-conditioning, Gerty worked until the very last moment before going to the maternity hospital. Tom had an impressive career in his own right later and made important contributions to biochemical research by founding the Sigma Chemical Company, which manufactured and distributed an enormous catalog of biochemical reagents. He was a rather unruly child whom the research fellows feared to babysit. He describes himself as a hothead and recalcitrant, with an "in your face" personality. He knew his parents loved each other despite their different personalities. He noted that they became American citizens after witnessing the events of World War II. He described his father as shy and cerebral and his mother as more emotional. Politically, Carl was liberal and Gerty was conservative. Tom felt that he was more like his mother in temperament. His early education was at a private nursery school and then a private middle and secondary school. He went to the University of Wisconsin in 1954 and wanted to be career army officer and was headed to West Point, but was dissuaded by the idiosyncratic professor Paul Link. Link's personality and accomplishments are described in detail in chapter 26. In 1959, Tom went to Washington University for a PhD in chemistry under David Lipkin to study the intramolecular reactions of nucleotides, including cyclic AMP.

He decided not to go to Harvard because the academic life did not appeal to him. Instead, he called up the personnel department of the Sigma Chemical Company in St. Louis in 1970 and was hired by them. He was assigned to prepare ATP from horse muscle and then NAD and NADP from yeast and was impressed by the massive amounts of starting material required. One of the early hires of the company was Dan Broida who, according to the catalog, would answer queries about the products at any time of the day and from any country. This was true, but Broida tended to be grumpy if the call came through in the very early morning. One of the queries was from an investigator who couldn't dissolve a reagent. It turned out he was trying to dissolve the packing material. When the company merged with the Aldrich Chemical Company in 1975, Tom became a vice president and then president in 1980. He was chief executive officer from 1983 to 1999 and chairman in 1991, retiring at the end of 2000. From a base of $76 million in sales, 1,400 employees, and 24,000 products in 1980, Tom Cori guided the company to become a world leader in providing high-quality products for use in life science research and disease diagnosis. At the time he retired, there were 5,800 employees, 85,000 products, and sales exceeding $1 billion. Thus, while his parents were noted for their advances in biochemical research, Tom was a great facilitator of such research through his company's provision of essential reagents.

Award of Nobel Prizes and Studies of Glycolytic Enzymes

In 1947, the Coris were awarded the Nobel Prize in Physiology or Medicine, sharing it with the Argentinean physiologist Bernardo Houssay. The Coris received it on the basis of their discovery of the enzymatic breakdown of glycogen, whereas Houssay was awarded it because of his work on the role of the pituitary gland in controlling carbohydrate metabolism. Houssay's star pupil, Luis Leloir, later joined the Cori laboratory and received the Nobel Prize in 1970. His career is described in chapter 6. Unexpectedly, Gerty fell ill while climbing Snowmass, a mountain near Aspen in Colorado, and, a few weeks before the Coris were to depart for Stockholm came the terrible news that she was suffering from an incurable form of anemia (myelosclerosis). Despite her illness, she attended the Nobel ceremony and as many of the attendant events as possible. As related in chapter 18, she bore her illness with great fortitude, but eventually succumbed in October 1957.

The Coris' work on phosphorylase, including its crystallization and detailed characterization, has been described in chapter 5, but they extended their work to other enzymes involved in glycolysis, for example, hexokinase and glyceraldehyde-3-phosphate dehydrogenase. Their studies with the latter enzyme are principally described in chapter 9. Hexokinase was purified from yeast and its properties characterized. Another project was to explore the mechanisms of two other enzymes. The first was phosphoglucomutase, and the researchers involved, besides Cori, were Earl Sutherland, Mildred Cohn, and Theodore Posternak, a fellow from Lausanne, Switzerland. The project started when Victor Najjar, the subject of chapter 10, found a preparation of glucose-1-phosphate that couldn't be converted to glucose-6- phosphate by the enzyme. Further work showed a great variation in the ability of various preparations of glucose-1-phosphate to act as substrates for the enzyme, and they attributed this to the variable presence of an activator of the enzyme. The explanation came when Luis Leloir (see chapter 6) suggested that the activator for this reversible reaction was glucose-1,6-bisphosphate. Posternak synthesized this compound, and, with Mildred Cohn playing a key role, glucose-1-phosphate labeled with either ^{32}P or ^{14}C was used to show that the mechanism proposed by Leloir was correct. This was that the enzyme catalyzes the transfer of the phosphate from position 1 of glucose-1,6-bisphosphate to position 6 of glucose-1-phosphate. In this way, the diphosphate becomes glucose-6-phosphate, and the diphosphate is regenerated from glucose-1-phosphate. They recognized this as a unique reaction mechanism for an enzyme. A similar mechanism was found by them for the glycolytic enzyme phosphoglycerate mutase. This catalyzes reversible conversion of 3-phosphoglycerate to 2-phosphoglycerate with 2,3-phosphoglycerate as the cofactor.

Carl continued his enzymological studies with researches on phosphorylase in collaboration with Neil Madsen. Madsen was a Canadian who obtained a BSc and an MSc from the University of Alberta. He then spent two years with Carl Cori earning his PhD in 1955. His research involved treating phosphorylase *a* with the sulfhydryl reagent *p*-chloromercuribenzoate, which resulted in its cleavage into four identical inactive monomers. These results indicated that the enzyme required intact thiol groups for activity. Reactivation and reassociation could be induced by addition of cysteine. Madsen and Cori also studied the binding of glycogen and AMP to phosphorylase using ultracentrifugation. Madsen then spent a postdoctoral period with Hans Krebs at Oxford. His project there was carried out with Hans Kornberg, who was one of Krebs' star pupils and who later succeeded Sir Frank Young as the Sir William Dunn Professor of Biochemistry at Cambridge University. Madsen's project involved studying the metabolism of a pseudomonad bacterium, which led to the recognition that there was a second point of entry of acetate into the citric acid cycle. This became known as the "glyoxylate bypass" and launched Kornberg's career. In 1962, Madsen returned to the Biochemistry Department at his alma mater, where his research returned to a focus on phosphorylase whose crystal structure he determined in association with Robert Fletterick and others. He carried out structure-function studies to elucidate the site and mechanism of catalysis, the mechanism of activation by AMP, and the structural changes induced by phosphorylation.

Carl also collaborated with Gerty, Joseph Larner, and Barbara Illingworth in studies of the enzymes involved in glycogen storage diseases. These are described at length in chapters 18, 19 and 20. His interests also turned to the hormonal regulation of muscle carbohydrate metabolism (chapter 15). In 1956, a special issue of *Biochimica et Biophysica Acta* was published. It was titled "Enzymes and Metabolism: A Collection of Papers Dedicated to Carl F. and Gerty T. Cori on Occasion of their 60th Birthday." The introduction was by Bernardo Houssay, who described their careers and many significant accomplishments. He pointed to their imagination, experimental rigor, and critical judgment. He said that their vast general culture and thorough training in medicine, physiology, and pharmacology gave them an integrated view of chemical changes in vitro and in isolated tissues and intact animals under normal and pathological conditions. He also commented on their personal qualities, which won them the friendship and admiration of all who knew them. Houssay's introduction was followed by scientific articles by the six Nobel Laureates who worked in their department, plus twenty-six other distinguished scientists who spent time there.

References

Berger, L., M. W. Slein, S. P. Colowick, and C. F. Cori. 1946. "Isolation of Hexokinase from Baker's Yeast." *J. Gen. Physiol.* 29: 379–91.

Cohn, M. 1992. "Carl Ferdinand Cori 1896–1984." *Biogr. Mem. Natl. Acad. Sci. U.S.A.* 61: 79–109.

Cori, C. F., and N. B. Madsen. 1956. "The Interaction of Muscle Phosphorylase with *p*-Chloromercuribenzoate." *J. Biol. Chem.* 223: 1055–65.

Cori, T. Personal Communication.

Fletterick, R. J., and N. B. Madsen. 1980. "The Structures and Related Functions of Phosphorylase a." *Ann. Rev. Biochem.* 49: 31–61.

Kornberg, A. 2001. "Remembering Our Teachers." *J. Biol. Chem.* 276: 3–11.

Kornberg, H. L. 2003. "Memoirs of a Biochemical Hod Carrier." *J. Biol. Chem.* 278: 9993–10001.

Madsen, N. B., and C. F. Cori. 1957. "The Binding of Adenylic Acid by Muscle Phosphorylase." *J. Biol. Chem.* 224: 899–980.

Randle. P. J. 1986. "Carl Ferdinand Cori 5 December 1896–20 October 1984." *Biogr. Mem. Natl. Acad. Sci. U.S.A.* 61: 67–95.

Sutherland, E. W., M. Cohn, T. Posternak, and C. F. Cori. 1949. "The Mechanism of the Phosphoglucomutase Reaction." *J. Biol. Chem.* 180: 825–37.

Sutherland, E. W., T. Posternak, and C. F. Cori. 1949. "The Mechanism of the Phosphoglyceric Mutase Reaction." *J. Biol. Chem.* 181: 153–59.

9

Sidney Velick

Modest Enzymologist

Joining the Cori Laboratory

Sidney Frederick Velick was born in Detroit in 1913 and attended Detroit City College, now Wayne State University. His father was a lawyer who later joined the family scrap iron business. His father was killed during a robbery, and the scrap iron business failed, which created great difficulties for the family. They were forced to move to Oklahoma to live with an uncle and then back to Detroit. In spite of these problems, Velick did well in high school and enjoyed writing and reading classics such as Tolstoy's *War and Peace*. At college, he was editor of the feature page of the college newspaper. He felt his career would be in writing, but he classified himself as a pre-medical student. He did well in chemistry and biology, but received a C grade in Comparative Anatomy. This was due to the anti-Semitism of the instructor, who stated that he did not want to encourage the progress of another Jewish abortionist! Velick graduated with a BS in chemistry in 1935 and then obtained a PhD in biochemistry from the University of Michigan in 1938. He did postdoctoral research at Johns Hopkins under Robert Hegner, a parasitologist who wanted to employ a biophysicist. Velick worked at the School of Public Health at Hopkins, studying why malarial parasites (merezoites) preferentially invaded reticulated red cells. He also spent time at the biological laboratory at Cold Spring Harbor on Long Island, and this and his work at Hopkins resulted in two papers on red cells and the malarial parasite. When his preceptor at Hopkins became ill with cancer, Velick went to Yale to work on the lipids of *Phytomonas tumefaciens*, the bacterium that causes crown gall disease in plants.

He joined the faculty in the Cori department in 1945, accepting an offer from Carl Cori to be an assistant professor in the Department of Biochemistry at Washington University. Velick's time in St. Louis proved to be the most productive in his career. At that time, nothing was known about the amino acid sequences of proteins, let alone their amino acid composition. Velick determined the amino acid compositions of the glycolytic enzymes aldolase and glyceraldehyde-3-phosphate dehydrogenase, which required their purification to a high degree. However, he was more interested in determining their turnover, and, using isotopic techniques, he showed that they were synthesized from the same precursors but turned over at different rates. He worked on many other enzymes, but his primary interest was glyceraldehyde-3-phosphate dehydrogenase. In particular, he found that the enzyme contained tightly bound nicotinamide adenine dinucleotide (NAD), which is its cofactor. He also explored its mechanism of action, with emphasis on the role of the sulfhydryl groups, the conformation of NADH as bound to the enzyme, and the role of energy transfer to a tryptophan. Much of this work was done with Jane Harting and was the basis for her PhD degree. In 1953, Harting married Charles R. (Rollo) Park, another member of the Cori department, and, as described in greater detail in their respective chapters, they both became faculty members at Vanderbilt Medical School.

Careers of Murray Heimberg and Philip Strittmatter

One of the fellows who worked with Velick was Murray Heimberg. Heimberg was born in 1925 in Brooklyn, New York, and was drafted into the U.S. Army during World War II. He served as an infantryman in various parts of Europe, but was captured by the Germans near Aachen in November 1944 and spent the rest of the war in a prisoner of war camp. He described the amenities as poor and the hot summer hard to endure, with coal dust pouring through the windows. As a result of the poor nutrition while he was held captive, he lost half his body weight. With the end of the war in Europe, Heimberg returned to the United States and went to Cornell University as an undergraduate. He then went to Duke University for a PhD under the renowned biochemist Philip Handler, who was blessed with a photographic memory, excellent judgment, and an incomparable mastery of language. Handler later became president of the American Society of Biologists and president of the National Academy of Sciences, and received the National Medal of Science.

Handler assigned a junior graduate student, Irvin Fridovich, to work with Heimberg; their project was the enzymatic oxidation of sulfite. Fridovich remained at Duke, rising to become James B. Duke Professor of Biochemistry and an expert on the enzyme superoxide dismutase, which removes damaging superoxide from cells. Heimberg obtained his PhD in 1952 and considered doing a postdoctoral with either Vincent du Vigneaud or Carl Cori, but decided to go to the Cori department where he was assigned to work with Velick. As noted earlier, Velick's specialty was enzymology, and Heimberg's project was to study the synthesis of aldolase and phosphorylase in rabbits.

In 1954, Charles Park recruited Heimberg to Vanderbilt, and his first project was to work with Jane Park on the effects of thyroxine and triiodothyronine on glycolysis in ascites tumor cells. He then worked with H. C. (Ray) Meng, who was an original member of the Park department and whose specialty was lipid metabolism. Heimberg took the research in a different direction by developing an isolated rat liver perfusion system. This was then used to study various aspects of hepatic lipid metabolism. The approach enabled the hepatic effects of factors such as diabetes and hormones to be studied without the complications of effects in peripheral tissues. The approach led to an impressive series of studies of the regulation of the hepatic uptake and release of triacylglycerols (triglycerides), which are the major lipids in VLDL (very low density lipoproteins). Heimberg showed that they could be taken up by the liver and that this was increased by fasting and diabetes.

Heimberg then decided to get an MD at Vanderbilt Medical School. On graduation in 1959, he joined the Pharmacology Department there and continued his researches using the perfused liver preparation. The studies were extended to the effects of perfusion with free fatty acids (FFA) on triacylglycerol synthesis and ketone body production in livers from normal and diabetic rats. He also examined the effects of glucagon on the output of triacylglycerol and ketone body production. In 1974, he became chair of pharmacology at the University of Missouri, and in 1981, he was recruited to the same position at the University of Tennessee. Heimberg continued his work on liver lipid metabolism, looking at the effects of thyroid status and diet.

Another notable biochemist who worked with Velick is Philipp Strittmatter, who was born in Philadelphia in 1928. He obtained his PhD under Eric Ball at Harvard University, studying formaldehyde dehydrogenase, and in 1955 began a postdoctoral period in the Cori department, where he worked exclusively with Velick to study microsomal cytochrome b5 and its NAD-dependent reductase. He then began working independently on this enzyme system in 1958, and it became the major topic for the rest of his career. Strittmatter was interested in the transfer of hydrogen from NADH to the reductase and the amino acid sequence of cytochrome b5. In 1969, he was appointed chairman of biochemistry at the new University of Connecticut Medical School at Farmington. He continued working on cytochrome b5 and its reductase, with a focus on their binding to membranes and phospholipid vesicles, but also started working on microsomal stearyl-CoA desaturase. He retired from the chairmanship in 1994.

Another respected scientist who worked with Velick is Carl Frieden, whose career is documented in chapter 26. Frieden says that Velick fundamentally left him alone to study cytochrome c reductase.

Move to the University of Utah

Velick became interested in using fluorescence as a means to understand enzyme mechanisms, including measuring tryptophan fluorescence to determine ligand binding. His later work with Strittmatter, as described above, involved studies of microsomal cytochrome b and a microsomal cytochrome reductase. The latter was purified and characterized with a focus on the oxidation-reduction stoichiometry, in keeping with Velick's interest in biological oxidations. He also studied the catalytic mechanisms of some enzymes involved in amino acid metabolism. In 1964, Velick moved to Salt Lake City to become the chairman of biochemistry at the University of Utah School of Medicine and led the department until 1978. He continued his interest in glyceraldehyde-3-phosphate dehydrogenase with a further focus on its kinetic mechanism and the thermodynamics of NAD binding. The binding studies were extensive and included equilibrium and calorimetric analyses and measurements of protein hydration. Velick was a modest person and felt that he should not claim authorship on any paper unless he had a significant role in the research. Unfortunately, this is not always the case for laboratory directors.

At Utah, Velick had the burden of giving all the biochemistry lectures to the first-year medical students, and he did not receive all the space he had been offered. Because of space and budget limitations, the department remained small. However, Velick was able to spend enjoyable sabbaticals with Manfred Eigen in Göttingen and Ernst Helmreich in Würzburg. The careers of these scientists are described in chapter 25. Velick was treated with great fondness by the departmental members. He encouraged research in the department, but had a small laboratory himself.

When he retired from the chair and became an emeritus professor, he pursued research again as a member of the Genetics Group in the Department of Biology at the University of Utah, which was sponsored by the Howard Hughes Medical Institute. He served as executive editor of *Archives of Biochemistry and Biophysics* and received the Alexander von Humboldt Senior Award from West Germany and a Distinguished Research Award from the University of Utah. In 1981, he was elected to the National Academy of Sciences. Velick loved the outdoors and was an avid skier and hiker until well into his eighties; he also enjoyed rafting and canoeing on the rivers of Utah. He cofounded with his wife the Chamber Music Society of Salt Lake City and also the Utah Alliance for the Mentally Ill. Their daughter suffered from mental illness, and they were strong advocates for reforms in the treatment of this type of illness. Velick died in Salt Lake City on December 29, 2007, at the age of ninety-four.

References

Frieden, C. 2010. "Sidney Frederick Velick." *Biogr. Mem. Natl. Acad. Sci. U.S.A.* 1–14.
Harting, J., and S. F. Velick. 1954. "Transfer reactions of Acetylphosphate Catalyzed by Glyceraldehyde-3 Phosphate Dehydrogenase." *J. Biol. Chem.* 207: 867–78.
Heimberg, M., Meng, H. C., and C. R. Park. 1958. "Effect of Sex, Fasting and Alloxan Diabetes on the Uptake of Neutral Fat by Isolated Perfused Rat Liver." *Am. J. Physiol.* 195: 673–77.

Heimberg, M., J. H. Park, A. Isaacs, and R. Pitt-Rivers. 1955. "The Effect of Acetic Acid Analogs of Thyroxine on Glycolysis in Ascites Tumor Cells in Vitro." *Endocrinology* 57: 756–58.

Heimberg, M., and S. F. Velick. 1954. "The Synthesis of Aldolase and Glyceraldehyde-3-Phosphate Dehydrogenase in the Rabbit." *J. Biol. Chem.* 208: 725–30.

Heimberg, M., I. Weinstein, and M. Kohout. 1969. "The Effects of Glucagon, Dibutyryl Cyclic Adenosine 3′,5′-Monophosphate and Concentration of Free Fatty Acid on Hepatic Lipid Metabolism." *J. Biol. Chem.* 244: 5131–39.

Olubadewo, J., D. W. Morgan, and M. Heimberg. 1983. "Effects of Triiodothyronine on Biosynthesis and Secretion of Triglyceride by Livers Perfused in vitro with [H^3]Oleate and [C^{14}]Glycerol." *J. Biol. Chem.* 258: 938–45.

Simpson, M. V., and S. F. Velick. 1954. "The Synthesis of Aldolase and Glyceraldehyde-3-Phosphate Dehydrogenase in the Rabbit." *J. Biol. Chem.* 208: 61–71.

Soler-Argilaga, C., H. G. Wilcox, and M. Heimberg. 1976. "The Effect of Sex on the Quantity and Properties of the Very Low Density Lipoprotein Secreted by the Liver in vitro." *J. Lipid Res.* 17: 139–45.

Strittmatter, P., and S. F. Velick. 1956. "The Isolation and Properties of Microsomal Cytochrome." *J. Biol. Chem.* 221: 253–64.

Strittmatter, P., and S. F. Velick. 1957. "The Purification and Properties of Microsomal Cytochrome Reductase." *J. Biol. Chem.* 228: 785–99.

Strittmatter, S. M. Personal Communication.

Van Harken, D. R., C. W. Dixon, and M. Heimberg. 1969. "Hepatic Lipid Metabolism in Experimental Diabetes. V. The Effect of Concentrationof Oleate on Metabolism of Triglycerides and Ketogenesis." *J. Biol. Chem.* 244: 2278–85.

Velick, S. F. 1953. "Coenzyme Binding and the Thiol Groups of Glyceraldehyde-3-Phosphate Dehydrogenase." *J. Biol. Chem.* 203: 563–73.

Velick, S. F., J. P. Baggott, and J. M. Sturtevant. 1971. "Thermodynamics of Nicotinamide Adenine Dinucleotide Addition to the Glyceraldehyde-3-Phosphate Dehydrognases of Yeast and Rabbit Skeletal Muscle." *Biochemistry* 10: 779–86.

Velick, S. F., J. R. Hayes Jr., and J. Harting. 1953. "The Binding of Diphosphopyridine Nucleotide by Glyceraldehyde-3-Phosphate Dehydrogenase." *J. Biol. Chem.* 203: 527–44.

Velick, S. F., and E. Ronzoni. 1948. "The Amino Aid Composition of Aldolase and Glyceraldehyde-3-Phosphate Dehydrogenase." *J. Biol. Chem.* 173: 627–39.

10

Victor Najjar

Pediatrician and Immunochemist

Career in Pediatrics at Johns Hopkins

Victor Assad Najjar was one of the earliest postdoctoral fellows to work with the Coris in St. Louis. He possessed great personal charm and experimental ability. He was born in Beirut, Lebanon, on April 15, 1914, and graduated MD from the Medical School of the American University of Beirut in 1935. After an internship in pediatrics there, he went with his brother to Baghdad to work in public health at the Teachers College Hospital. In 1938, he went to Johns Hopkins Hospital to do a residency in pediatrics under the direction of the chairman, Edwards Park, the father of Charles R. Park, who also later joined the Cori laboratory. Najjar was impressed by the collaborative atmosphere and networking in the Pediatric Department under Edwards Park's leadership. Some of Najjar's clinical studies at Hopkins included the demonstration that the bacterial synthesis of thiamine in the intestinal tract supplies part of the daily requirement for this vitamin, and that N-methyl nicotinamide is a major metabolite of nicotinamide.

In 1944, Najjar went to the Harriet Lane Home as an assistant professor and director of the outpatient department. The Home was affiliated with Johns Hopkins under the direction of Emmett Holt Jr., who was a leader and outstanding contributor to the field of clinical nutrition. Holt's father, Emmett Holt Sr., was also a leader in American pediatrics and a founding member and president of the American Pediatric Society.

During his time at Hopkins, Najjar became great friends with Charles Park. Najjar served as Park's best man at his wedding to Jane Harting and also hosted the wedding reception.

In 1946, Najjar received a National Research Council Award to work in the Cori laboratory His project was the isolation and purification of phosphoglucomutase, which was previously identified by the Coris, working with Sidney Colowick. Najjar succeeded in crystallizing the enzyme and characterizing its activity. This was a remarkable achievement since Najjar had no previous experience in this area. He also worked with Edwin Krebs to develop antibodies to glyceraldehyde-3-phosphate dehydrogenase and test their effects on the activity of the enzyme. It was during this time that he married his wife Matilda, who was doing research at Barnes Hospital and shared a laboratory with Najjar.

In 1948, Najjar went to work with Ernest Gale, a well-known microbiologist in the Department of Biochemistry at Cambridge University. Gale was notable because he emphasized the chemical and enzymatic basis of microbial activity at a time when this was ill-defined. He also explored the molecular basis for the actions of antibiotics. In other words, he was ahead of his time, and Najjar's choice to work with Gale was a smart one.

Najjar then returned to Johns Hopkins as an associate professor of pediatrics. His most famous accomplishment there was to recognize a new form of congenital non-hemolytic jaundice. This was done in association with John F. Crigler and is known as the Crigler-Najjar Syndrome. This illness is now recognized to be due to a deficiency of bilirubin UDP glucuronyl transferase, which is the enzyme that converts bilirubin to bilirubin glucuronide, thus allowing it to be excreted in the bile. Najjar enjoyed working on biochemical topics and interacting with Sidney Colowick and Nathan Kaplan at the associated McCollum-Pratt Institute. He continued to work on the enzymatic mechanism of phosphoglucomutase and the kinetics of hexokinase. The significance of the work with phosphoglucomutase was that the enzyme actually participates as an acceptor and a donor of the phosphate group whose transfer it catalyzes. This was the first demonstration that an enzyme could participate as a reactant in a group transfer process.

Another accomplishment was his formulation of the mechanism of the antigen-antibody reaction. This describes that when an antigen is introduced into the body, it stimulates the formation of specific antibodies to special sites on its surface. When it is introduced again, it reacts rapidly with the earlier antibodies to form an antibody-antigen complex. In this complex, the configuration of the antigen and antibody are mutually altered, and the complex behaves as a new antigen. Specific antibodies are then formed against the new sites of altered configuration.

Move to Vanderbilt and Tufts Universities

In 1957, Najjar was appointed chairman of the Microbiology Department at Vanderbilt University (Figure 13). Charles Park was instrumental in that appointment. There was some limited resistance from the more conservative chairmen because of Najjar's Arabic heritage, but this subsided. Najjar brought to the chairmanship much vigor and enthusiasm for research. His strengths included an ability to facilitate research and to develop scientific relationships. He had a knack for simplifying complex events and reducing things to simple concepts. This ability to clarify issues meant that he was an effective lecturer. He began to focus more on immunochemistry, and some of his achievements included discovering specific γ-globulins, which bind respectively to red and white blood cells and play a role in their survival and function. He also continued working on phosphoglucomutase, showing that the substrate was activated by Mg^{2+} ions through the formation of a metallosubstrate complex. This was one of the first concepts of the role of metallosubstrates in metal-catalyzed reactions. Another finding was the identification of the amino acid sequence to which phosphate is covalently linked during the phosphoglucomutase reaction. Najjar recruited Sidney Colowick to the department as well as a cadre of younger scientists who expanded the research program, including James T. Park, who subsequently went to Harvard to work with Jack Strominger. Together, Park and Strominger made a major breakthrough, namely the elucidation of the mechanism of action of penicillin. Charles Park and Najjar were instrumental in bringing Earl Sutherland to Vanderbilt.

In 1968, the opportunity to create a Division of Protein Chemistry in the Department of Molecular Biology and Microbiology at Tufts University School of Medicine in Boston presented itself, and Najjar accepted. This gave him the opportunity to devote himself to research without the concerns of a chairmanship. At Tufts, he devoted himself principally to a peptide he had discovered at Vanderbilt, which he called Vankinin. When he moved to Tufts, he renamed it Tuftsin. This was found during studies of leukokinin, a γ-globulin fraction essential for maximal stimulation of the phagocytic activity of neutrophils. The whole effect was later ascribed to a single peptide fragment liberated by a specific enzyme on the outer membrane of the neutrophil. The significance of Tuftsin was indicated when it was found to be absent or in mutant forms in several patients with recurring infections. Isolation of Tuftsin showed it to be a tetrapeptide (L-threonyl-L-lysyl-L-prolyl-L-arginine), and in vitro studies showed that it stimulated all functions of phagocytic cells including phagocytosis, pinocytosis, motility, immunogenic activity, and bactericidal activity.

Tuftsin formation was shown to involve two different enzymes, and a receptor for it was isolated and its subunit structure defined. Other effects of Tuftsin were explored, namely its induction of tumor necrosis activity and its stimulation of the growth of HL60 promyelocytic cells, which also served as a source for the Tuftsin receptor. The peptide was shown to reside in the Fc domain of the heavy chain of immunoglobulin G. It was synthesized chemically, and this allowed extensive studies of its structure-function relationships. In a different area, Najjar studied the mechanism by which adenylyl cyclase is activated by fluoride and hormones, although the proposal that the enzyme was regulated by phosphorylation was superseded by the discovery that G proteins were involved.

In 1978, Najjar was appointed American Cancer Society Research Professor of Molecular Biology and Microbiology at Tufts, where he remained until he retired in 1984. He returned in the early 1990s to Nashville where his daughter Jennifer was a pediatric endocrinologist and a member of the Department of Pediatrics at

Vanderbilt. During Najjar's prolonged battle with Parkinson's disease, he had the support of his family as well as that of his long-time friend, Charles Park. Najjar died in Nashville on December 6, 2002.

References

Bump, N. J., J. Lee, M. Wleklik, J. Reichler, and V. A. Najjar. 1986. "Isolation and Subunit Composition of Tuftsin Receptor." *Proc. Natl. Acad. Sci. U.S.A.* 83: 7187–91.

Constantopoulos, A., V. A. Najjar, and J. W. Smith. 1972. "Tuftsin Deficiency: A New Syndrome with Defective Phagocytosis." *J. Pediatr.* 80: 564–72.

Crigler, J. F. Jr., and V. A. Najjar. 1952. "Congenital Familial Jaundice with Kernicturus." *Am. J. Dis. Child.* 83: 259–60.

Fidalgo, B. V., and V. A. Najjar. 1967. "The Physiological Role of the Lymphoid System. 3. Leucophilic Gamma-Globulin and the Phagocytic Activity of the Polymorphonuclear Leucocyte." *Proc. Natl. Acad. Sci. U.S.A.* 57: 957–64.

Fidalgo, B. V., V. A. Najjar, C. F. Zukoski, and Y. Katayama. 1967. "The Physiological Role of the Lymphoid System. II. Erythrophilic Gamma Globulin and Survival of the Erythrocyte." *Proc. Natl. Acad. Sci. U.S.A.* 57: 665–72.

Fridkin, M., and V. A. Najjar. 1989. "Tuftsin: Its Chemistry, Biology and Clinical Potential." *Crit. Rev. Biochem. Mol. Biol.* 24: 1–40.

Harshman, S., J. P. Robinson, V. Bocchini, and V. A. Najjar. 1965. "Activation of Phosphoglucomutase." *Biochemistry* 4: 396–400.

Harshman, S,. H. R. Six, and V. A. Najjar. 1969. "The Sequence of a Phosphorylated Hexadecapeptide from Rabbit Muscle Phosphoglucomutase." *Biochemistry* 8: 3417–23.

Najjar, J, and M. Najjar. Personal Communication.

Najjar, V. A. 1948. "The Isolation and Properties of Phosphoglucomutase." *J. Biol. Chem.* 175: 281–90.

Najjar, V. A., and J. Fisher. 1955. "Mechanism of Antibody-Antigen Reaction." *Science* 122: 1272–73.

Najjar, V. A., and K. Nishioka. 1970. "Tuftsin: A Natural Phagocytosis Stimulating Peptide." *Nature* 228: 672–73.

Najjar, V. A., and M. E. Pullman. 1954. "The Occurrence of a Group Transfer Involving Enzyme (Phosphoglucomutase) and Substrate." *Science* 119: 631–34.

Najjar, V. A., and V. White. 1944. "F2 and N1-Methylnicotinamide." *Science* 100: 247–48.

Nishioka, K., P. S. Sato, A. Constantopoulos, and V. A. Najjar. 1973. "The Chemical Synthesis of the Phagocytosis-Stimulating Tetrapeptide Tuftsin (Thr-Lys-Pro-Arg) and Its Biological Properties." *Biochim. Biophys. Acta* 17: 230–37.

Robinson, J. P., S. Harshman, and V. A. Najjar. 1965. "Catalytic Properties of Activated and Nonactivated Phosphoglucomutase." *Biochemistry* 4: 401–05.

11

Edwin Krebs

Accidental Biochemist

Growing Up in the Great Depression

Like Luis Leloir, Edwin Gerhard Krebs (Figure 14) spent four years in the Cori laboratory. He was born on June 6, 1918, in Lansing, Iowa, the third of four children of Helen Stegeman Krebs and William Carl Krebs whose families had settled in Wisconsin after emigrating from Germany. His father was a minister who was trained at a Moravian seminary in Pennsylvania, but switched to Presbyterianism. His mother was a schoolteacher prior to her marriage. In 1919, the family moved to Newton, Illinois, and to Greenville, Illinois, in 1925. Krebs attended grade school and one year of high school at Greenville, which he described as a small college town in a pleasant part of southern Illinois.

Krebs was fascinated by the Civil War and read every book he could find on the subject. He had no particular interest in science apart from building a shortwave radio, but he did make gunpowder. He loved to fish and swim in the local streams, and he joined the Boy Scouts, but dropped out when they started having meetings in churches. He had happy experiences in elementary school and was encouraged by his parents to do as well as possible. His father died unexpectedly when he was fifteen years old, and this affected him deeply. His mother was devastated but soon began picking up the pieces and planning for the future, especially the education of her children. She decided to move to Urbana, Illinois, where two of her sons were students at the University of Illinois and where the tuition was only $35 per semester. High school for Krebs in Urbana was different from that in Greenville because of the excellent faculty, the influence of the nearby university, and the high percentage of college-bound students.

His father's death caused him to think about his career, and this preoccupied his mind for several years. It was the time of the Great Depression, and his family's financial situation was precarious. To help out, he tried door-to-door sales but was remarkably unsuccessful. He entered the University of Illinois with the idea that any courses he took should lead to a job, but the problem was that he couldn't decide what he wanted to do. He felt the business world was not for him and that law was boring. That left medicine and science as the best possibilities, and his courses during the first two years were suitable for both. He discovered a plan called individual curriculum, which gave him courses in chemistry, physics, mathematics, and enough biology to qualify for medical school. What he lost out on was languages and humanities, something he always regretted. He was able to do some research in organic chemistry while working as a dishwasher under a New Deal program.

Medical School and the U.S. Navy

Krebs was offered a scholarship at Washington University and met with the dean of the medical school, Philip Shaffer, who was also chair of the Department of Biological Chemistry. Shaffer was unique as dean since he was a PhD and not an MD, but this was because the medical school was highly research-oriented. He made it clear to Krebs that, in addition to training students for the practice of medicine, the school was dedicated to training students for medical research. Krebs took the usual first-year courses and became acquainted with members of the Department of Pharmacology, including both Coris, Arda Green, Sidney Colowick, and Earl Sutherland. He was invited to attend their weekly seminars, and Arda Green showed him the first preparations of crystalline phosphorylase.

Krebs' career in research was cut short abruptly on December 7, 1941, with the involvement of the United States in World War II, when the medical students joined the army or navy as reserve officers. Krebs joined the Naval Reserve, which meant that he was exempted from active duty until he completed medical school. The school asked him to help teach biochemistry, but this meant that he missed out on one trimester of the regular course. As a result, Krebs did not participate in the surgical service, which was where students learned how to sew up wounds and put in sutures. This later caused some embarrassment when he was a medical officer aboard ship. Under wartime rules, graduates were required to take nine months of internship and eighteen months of assistant residency before going on active duty. Krebs took this training at Barnes Hospital in St. Louis and was impressed by some of the professors in internal medicine and hematology whom he describes as excellent teachers and wonderful human beings. In turn, Krebs impressed the faculty with his analysis of serum proteins using the newly acquired Tiselius apparatus. It was during this time that Krebs married Virginia Frech, a student at the Washington University School of Nursing.

Krebs spent a year in active duty in the Navy after World War II ended. As the sole medical officer on a ship in the Pacific, his lack of training in surgery caught up with him. But luckily, his chief pharmacist mate was able to teach him how to sew up cuts. One patient who had a severe facial laceration that required twenty sutures wrote to Krebs many years later and expressed his gratitude for the fine job he had done.

After his release from the navy, Krebs returned to St. Louis to continue his training in internal medicine, but he would have to wait two years for an opening in the residency program. He was advised by the chairman of medicine to obtain experience in biochemistry as an adjunct to a career in internal medicine. Carl Cori accepted him, and he began as a postdoctoral fellow in the fall of 1946.

Research in the Cori Laboratory

Krebs started in the Cori laboratory by checking the purity of two enzymes (glyceraldehyde-3-phosphate dehydrogenase and phosphorylase) that had been crystallized in the laboratory. Another project involved comparing the properties of the muscle and yeast enzymes and testing their cross reactivity to antibodies. In this, Krebs was joined by Victor Najjar, who is the subject of the previous chapter. A final project concerned the effects of protamine, AMP, and IMP on phosphorylase. At first, the Coris were hesitant to let him publish the findings, considering them to be "phenomenological"—this was before the important concepts of conformational changes in proteins and of allostery had been formulated. Krebs enjoyed his interactions with the many faculty members, postdoctoral fellows, and visiting scientists and noted that the equipment was a far cry from the sophisticated apparatus now available.

In his autobiography, Krebs described the Cori laboratory. He said he couldn't say enough about the benefit he gained from working with the Coris, but found it difficult to pinpoint what exactly made it such a good place. Both Coris set very high standards in research, and nothing was accepted as established until all possible evidence was brought to bear. Everyone worked hard, but nobody described it as drudgery, because it led to new and exciting results. They were given great freedom with respect to the selection of projects so long as those projects were appropriate for the laboratory. Krebs and his colleagues were encouraged to pursue problems that were new and original; the Coris had no respect for those who jumped on a bandwagon and did "me too" research. It was also essential that proper credit was given to other investigators for their contributions.

Access to either Carl or Gerty Cori was always available if you had something important to say or to show, but they were not to be interrupted to engage in idle chatter. When Krebs went to Carl to see about postdoctoral work after leaving the Navy, he felt that he should engage in the usual pleasantries, but Carl cut him short rather abruptly by asking: "Are you here just to pass the time of day, or do you want to talk business?" This was "classic" Cori, although Krebs noted that Carl did mellow with time.

Research at the University of Washington with Edmond Fischer

Krebs began to give serious thought about what to do after his two years in the Cori laboratory. A likely course was to return to residency training at Barnes Hospital. There were also positions for physicians with biochemical training, but there was a possible position in the Biochemistry Department in a new medical school at the University of Washington being established in Seattle. Krebs had visited there while in the navy and had been captivated by the climate and scenery. He was offered and accepted the job. He hated to give up internal medicine but felt the same about biochemistry. To be on the safe side, he obtained a license to practice medicine in the state of Washington. The first permanent chair of the department was Hans Neurath, who proved to be an excellent appointment. According to Krebs, Neurath was appropriately aggressive when it was warranted, but open and fair when dealing with people. Like the Coris, Neurath set a great example in research and other academic aspects. At that time, grants for research support from NIH were easily obtained—all one had to do was to write a letter! Krebs took an afternoon off to compose a letter, and a few months later the money arrived.

He started working on glyceraldehyde-3-phosphate dehydrogenase, and his first graduate student discovered that their enzyme was identical to a protein that had been crystallized at the Rockefeller Institute. They discovered that it existed in multiple forms, but this attracted little attention because this was an era before the existence of isozymes was recognized. Krebs' greatest contribution came when he and Edmond Fischer joined forces to study the regulation of phosphorylase.

Fischer was born in Shanghai, where his father owned a newspaper that was published in French. He had attended a boarding school in Switzerland and, in high school, he was admitted to the Geneva Conservatory of Music and was sufficiently talented to consider a career as a professional pianist. Fischer studied at the University of Geneva where he obtained a PhD in organic chemistry, studying α-amylase, an enzyme involved in the breakdown of starch and glycogen. He decided to go to CalTech, but he spent only a few months there because he received an invitation from Hans Neurath to join his department in Seattle. Six months after his arrival, Fischer started working with Krebs on phosphorylase. He had studied potato phosphorylase, whereas Krebs had worked on the muscle enzyme. As described earlier, it was known that the enzyme existed in two forms, a and b, and the Coris thought that the removal of a prosthetic group was responsible for the conversion of phosphorylase a to phosphorylase b. Since phosphorylase b required AMP for activity, it was thought that the prosthetic group was some form of AMP.

Krebs and Fischer decided to test this hypothesis, and their first foray met with surprise and disappointment. Although Krebs had isolated phosphorylase a several times while in the Cori laboratory, he was unable to do so in Seattle. All he could get from muscle was phosphorylase *b*. He and Fischer realized that they had altered one step, substituting a centrifugation step for a filtration step. When they changed to the Cori and Green procedure, they could obtain phosphorylase *a*. As described below, this is one example of many encountered in science, when pursuit of a failed or negative experiment can lead to a major finding. Krebs and Fischer found that aging of the muscle extracts prior to filtration did not yield phosphorylase *a*, and through an intuitive line of reasoning, they thought this was due to loss of ATP and that a kinase was involved in the conversion of phosphorylase *b* to phosphorylase *a*. They named this phosphorylase kinase.

It turned out that the role of the filter paper was to provide Ca^{2+} ions that were needed for the kinase reaction. But the situation became more complicated when it was discovered that the Ca^{2+} ions could also activate phosphorylase kinase by acting through a calcium-dependent protease now called calpain. Krebs and Fischer recognized that the conversion of phosphorylase *a* to phosphorylase *b* was performed by a protein phosphatase and not by an enzyme which removed a prosthetic group. Their findings were the first demonstration that protein phosphorylation was a dynamic process by which the activity of enzymes and other proteins could be controlled. This discovery led to the award of a Nobel Prize to both of them in 1992. They noted that similar findings had been obtained independently by Sutherland, Wosilait, and Rall working with liver phosphorylase. In the interest of historical accuracy, they pointed out that Eugene Kennedy had reported the phosphorylation of casein by a kinase as early as 1954.

Their research didn't stop there, and another major discovery was yet to be made. They were aware of Sutherland's findings that cyclic AMP was the mediator of the activation of phosphorylase by glucagon and epinephrine, but it was unclear whether this was the result of an effect on phosphorylase phosphatase or phosphorylase kinase. Krebs and Fischer found an effect of cyclic AMP in muscle extracts. They also found that phosphorylase kinase existed in a highly active phosphorylated form and a less active nonphosphorylated form. Furthermore, they found that cyclic AMP accelerated the activation of phosphorylase kinase in a reaction requiring ATP. However, they could not demonstrate a direct effect of cyclic AMP on phosphorylase kinase.

Discovery of Cyclic AMP-Dependent Protein Kinase

Finding the mechanism of the effect of cyclic AMP was not simple and took several years. The reason was that another protein kinase was contaminating their phosphorylase kinase preparations. An additional complication was that phosphorylase kinase could undergo autophosphorylation, that is, it could phosphorylate itself, and this caused activation. These problems were overcome by two fellows, Donal Walsh and John Perkins. The contaminating kinase was found to be the direct target of cyclic AMP and was eventually called cyclic AMP-dependent protein kinase (now usually abbreviated to PKA). The most exciting finding was that PKA could phosphorylate many other proteins and is now known to be involved in the regulation of a host of major physiological processes, for example, cardiac function and blood vessel contractility, fuel (glycogen and lipid) mobilization, gastrointestinal and urinary tract functions, and hormone and neurotransmitter release and actions. The discovery of PKA and its phosphorylation of phosphorylase kinase, which in turn phosphorylated phosphorylase, led to the concept of the control of enzymes and other proteins through a phosphorylation cascade. There are now numerous examples of control through this mechanism. Research on protein phosphorylation in the 1970s was so extensive that 5 percent of papers published by biological journals then were concerned with it.

Move to the University of California at Davis

In 1968, Krebs moved to the chair of biological chemistry at the medical school that was being established at Davis, California, near Sacramento. His reasons for moving were his interest in teaching medical students, his respect for Loren Carlson, the person organizing the basic sciences course, the presence of an active graduate program, and the attitude of Paul Stumpf, the chairman of biochemistry in the College of Agriculture, who was doing everything he could to help the new department get started. In addition to Walsh, Krebs now depended on his postdoctoral fellows, graduate students, and visiting scientists to keep his research going, and he was blessed with some outstanding people who went on to distinguished careers.

The projects undertaken by Krebs at Davis involved identifying hormone-sensitive lipase as a target of PKA. This was important because it explained how epinephrine mobilizes free fatty acids from fat stores in adipose tissue. Another major finding was that PKA was composed of two types of subunit—regulatory and catalytic—and that cyclic AMP caused the enzyme to dissociate with the release of free active catalytic subunit. Thus the way in which cyclic AMP activated the kinase was explained. Other work explored the subunit composition of phosphorylase kinase and the effects of PKA phosphorylation in phosphorylase kinase and glycogen synthase. Peptides were synthesized to explore the specificity of the kinase and to test if they had inhibitory activity. A novel area of research was begun by Jim Maller, who microinjected protein kinase subunits into *Xenopus laevis* oocytes to see effects on maturation.

Return to the University of Washington

During a sabbatical stay in Fischer's laboratory, Krebs was approached about returning to the University of Washington as chairman of pharmacology and as an investigator of the Howard Hughes Medical Institute. He agreed to return to Seattle and found his second stint as a chairman less demanding than the first because he had learned to delegate some of the administrative duties. One issue was that he wasn't a true pharmacologist, but this never bothered Carl Cori! The Hughes Institute was generous in remodeling the space, and their support allowed Krebs to engage in a broader range of topics. At the end of 1990, the Institute terminated its support ostensibly on the grounds of Krebs' age. This was handled in an ungracious and insensitive manner, and the Institute was roundly criticized. As a form of poetic justice, Krebs received the Nobel Prize two years later, but that did not prevent the Institute from celebrating the event. He was later made an emeritus investigator at the Institute.

Krebs' research at the University of Washington, in collaboration with Kenneth Walsh and Koiti Titani, involved determining the primary structure (amino acid sequence) of several proteins. These included the catalytic and regulatory subunits of PKA, the cyclic GMP-dependent protein kinase PKG (which was later found to mediate the effects of nitric oxide), myosin light chain kinase (involved in smooth muscle contraction), and subunits of casein kinase. In addition, structure-function studies were carried out, and the binding sites for ATP and substrates were determined. It was known that Ca^{2+}-calmodulin was a major regulator of myosin light chain kinase, and the binding site for this was determined.

An exciting development in the protein kinase field came with the finding by Tony Hunter at the Salk Institute that tyrosine residues on proteins could be phosphorylated. Up to this time, only serine and threonine residues were the known targets of protein kinases. Now tyrosine residues were recognized as the targets of $p60^{src}$, which is the transforming protein of the Rous sarcoma virus, indicating that the protein encodes a tyrosine kinase. Ray Erikson at Harvard had recognized earlier that the protein had kinase activity, but its special nature was not realized. The receptor for epidermal growth factor was also found by Stanley Cohen at Vanderbilt to have tyrosine kinase activity, and Ronald Kahn, working at the Joslin Diabetes Center in Boston, found that this was true for the insulin receptor. In 1986, Cohen and Rita Levi-Montalcini won the Nobel Prize for their work on growth factors. Since that time, many tyrosine kinases have been discovered that are either associated with receptors or are cytosolic.

Krebs turned his attention to the finding that tyrosine kinase activity could be coupled to serine/threonine kinase activity, and many observations indicated that this could be a common event. He focused on the activation of mitogen-activated protein kinase (MAPK) by growth factors and found that several protein kinases were involved. Thus, growth factor activation of its receptor tyrosine kinase leads to activation of Ras, a low molecular weight G protein. In turn, this activates Raf, a serine/threonine protein kinase, which phosphorylates and activates MAP kinase kinase (MAPKK), which then phosphorylates MAP kinase leading to its activation. The importance of this signal or phosphorylation cascade is that it is involved in the regulation of cell growth and can be deranged in certain cancers, for example when Ras is mutated to an oncogenic form. It is now recognized that many other signal cascades involving other protein kinases exist.

Soon after retirement, Krebs suffered a heart attack, which left him physically impaired and mainly confined to his house. He died on December 21, 2009, of heart failure. In his autobiography in the *Annual Reviews of Biochemistry*, he reflected on how phosphoproteins had been described fifty years ago in a biochemistry textbook—as just proteins in milk and egg yolk that were important in infant nutrition. He commented wryly that it is safe to say there is more to it than that! However, he noted that the authors of the article did suggest that the phosphorylation of proteins might confer new properties on them. He pointed out that phosphorylation of proteins is now the major means by which their functions are regulated. Although there are other types of reversible covalent modifications of proteins, none of them approach phosphorylation in terms of frequency of occurrence.

In addition to the Nobel Prize, Krebs received numerous other awards: the Lasker Award for Basic research, the Welch Award in Chemistry, the George W. Thorn Award, the FASEB 3M Award, and the American Heart Association's Research Achievement Award. He was elected to the National Academy of Sciences and the American Academy of Arts and Sciences. He was president of the American Society for Biological Chemistry in 1985 and was an associate editor of the *Journal of Biological Chemistry* from 1972 to 1993. Like Luis Leloir, he was rather humble and never boasted about his great achievements. He was extremely honest and honorable, and his experiences during the Great Depression made him careful about his money and that of the institutions and organizations that supported him. His quiet demeanor and concern for others earned him great respect from friends and colleagues. In his final words in his autobiography, he said that he became a biochemist by a circuitous route and was never completely sure that he was making the right decisions along the way. Even in a faculty position, he occasionally asked himself whether becoming a biochemist was the right choice. After more than fifty years in the field, he said that he was finally becoming convinced it was the right thing.

References

Ahn, N. G., R. Seger, R. L. Bratlien, C. D. Diltz, N. K. Tonks, and E. G. Krebs. 1991. "Multiple Components of an Epidermal Growth Factor-Stimulated Protein Kinase Cascade. In vitro Activation of a Myelin Basic Protein/Microtubule-Associated 2 Kinase." *J. Biol. Chem.* 266: 4220–27.
Brostrom, C. O., J. D. Corbin, C. A. King, and E. G. Krebs. 1971. "Interaction of the Subunits of Adenosine 3',5'-Cyclic Monophosphate-Dependent Protein Kinase of Muscle." *Proc. Natl. Acad. Sci. U.S.A.* 68: 2444–47.
Corbin, J. D., E. M. Reimann, D. A. Walsh, and E. G. Krebs. 1970. "Activation of Adipose Tissue Lipase by Skeletal Muscle Cyclic Adenosine 3',5'-Monophosphate-Stimulated Protein Kinase." *J. Biol. Chem.* 245: 4849–51.

Fischer, E. H., and E. G. Krebs. 1955. "Conversion of Phosphorylase *b* to Phosphorylase *a* in Muscle Extracts." *J. Biol. Chem.* 216: 121–32.

Graves, D. J., E. H. Fischer, and E. G. Krebs. 1960. "Specificity Studies on Muscle Phosphorylase Phosphatase." *J. Biol.Chem.* 235: 805–09.

Green, A. A., G. T. Cori., and C. F. Cori. 1942. "Crystalline Muscle Phosphorylase." *J. Biol. Chem.* 142. 447–48.

Hayakawa, T., J. P. Perkins, and E. G. Krebs. 1973. "Studies of the Subunit Structure of Rabbit Skeletal Muscle Phosphorylase Kinase." *Biochemistry* 12: 574–80.

Hunter, T. 1998. "The Croonian Lecture 1997. The Phosphorylation of Proteins on Tyrosine: Its Role in Cell Growth and Disease." *Phil. Trans. R. Soc. Lond. B.* 353: 583–605.

Hunter, T., and J. A. Cooper. 1985. "Protein-Tyrosine Kinases." *Ann. Rev. Biochem.* 54: 897–930.

Hunter, T., and B. M. Sefton. "Transforming Gene Product of Rous Sarcoma Virus Phosphorylates Tyrosine." 1980. *Proc. Natl. Acad. Sci. U.S.A.* 77: 1311–15.

Huston, R. B., and E. G. Krebs. 1968. "Activation of Skeletal Muscle Phosphorylase Kinase by Ca^{2+}. II Identification of the Kinase Activating Factor as a Proteolytic Enzyme." *Biochemistry* 7: 2116–22.

Kemp, B. E., E. Benjamini, and E. G. Krebs. 1976. "Synthetic Hexapeptide Substrates and Inhibitors of 3',5'-Cyclic AMP-Dependent Protein Kinase." *Proc. Natl. Acad. Sci. U.S.A.* 73: 1038–42.

Krebs, E. G. 1998. "An Accidental Biochemist." *Annu. Rev. Biochem.* 67: xiii–xxxii.

Krebs, E. G., D. J. Graves, and E. H. Fischer. 1959. "Factors Affecting the Activity of Muscle Phosphorylase Kinase." *J. Biol. Chem.* 234: 2867–73.

Krebs, E. G., and V. A. Najjar. 1948. "The Inhibition of Glyceraldehyde-3-Phosphate Dehydrogenase by Specific Antiserum." *J. Exp. Med.* 88 569–77.

Maller, J. L., and E. G. Krebs. 1980. "Regulation of Oocyte Maturation." *Curr. Top. Cell. Regul.* 16: 271–311.

Meyer, W. L., E. H. Fischer, and E. G. Krebs. 1964. "Activation of Skeletal Muscle Phosphorylase *b* Kinase." *Biochemistry* 3: 1033–39.

Pike, L. J., D. F. Bowen-Pope, R. Ross, and E. G. Krebs. 1983. "Characterization of Platelet-Derived Growth Factor-Stimulated Phosphorylation in Cell Membranes." *J. Biol. Chem.* 258: 9383–90.

Reimann, E. M., C. O. Brostrom, J. D. Corbin, C. A. King, and E. G. Krebs. 1971. "Separation of Regulatory and Catalytic Subunits of the Cyclic 3',5'-Adenosine Monophosphate-Dependent Protein Kinase of Muscle." *Biochem. Biophys. Res. Commun.* 42: 187–90.

Seger, R., and E. G. Krebs. 1995. "The MAPK Signaling Cascade." *FASEB J.* 9: 726–35.

Soderling, T. R., J. P. Hickenbottom, E. M. Reimann, F. L. Hunkeler, D. A. Walsh, and E. G. Krebs. 1970. "Inactivation of Glycogen Synthetase and Activation of Phosphorylase Kinase by Muscle Adenosine 3',5'-Monophosphate-Dependent Protein Kinases." *J. Biol. Chem.* 245: 6317–28.

Takio, K., D. K. Blumethal, K. A. Walsh, K. Titani, and E. G. Krebs. 1986. "Amino Acid Sequence of Rabbit Skeletal Muscle Myosin Light Chain Kinase." *Biochemistry* 24: 8049–57.

Takio, K., E. A. Kuenzel, K. A. Walsh, and E. G. Krebs. 1987. "Amino Acid Sequence of the Beta Subunit of Bovine Lung Case in Kinase II." *Proc. Natl. Acad. Sci. U.S.A.* 84: 4851–55.

Takio, K., S. B. Smith, E. G. Krebs, K. A. Walsh, and K. Titani. 1984. "Amino Acid Sequence of the Regulatory Subunit of Bovine Type II Adenosine Cyclic 3',5'-Phosphate-Dependent Protein Kinase." *Biochemistry* 18: 4200–06.

Takio. K., S. B. Smith, K. A. Walsh, E. G. Krebs, and K. Titani. 1983. "Amino Acid Sequence Around a 'Hinge' Region and Its 'Autophosphorylation' Site in Bovine Lung cGMP-Dependent Protein Kinase." *J. Biol. Chem.* 258: 5531–36.

Walsh, D. A., J. P. Perkins, and E. G. Krebs. 1968. "An Adenosine 3',5'-Monophosphate-Dependent Protein Kinase from Rabbit Skeletal Muscle." *J. Biol. Chem.* 243: 3763–65.

12

Mildred Cohn

Against All Odds

Early Career

Mildred Cohn was a diminutive woman who exhibited determination, tenacity, and a passion for science throughout her long career. As mentioned above, Mildred Cohn wrote a charming and informative biography of Carl Cori, and she described her scientific career in some delightful articles, which have formed the basis of this chapter. Her experiences demonstrate the challenges and obstacles faced by women wishing to be scientists at that time.

Mildred Cohn was born in New York City on July 12, 1913. Her mother and father had emigrated from the same Jewish community in Russia around 1905, although they did not know each other. Both sides of the family had a long rabbinical tradition, and her father had studied to be a rabbi but ended up as an inventor. He left rabbinical school to work in a tailor shop, where he invented a machine for cutting cloth accurately. Her father indoctrinated in her the belief that she could achieve anything she chose, but not without difficulty since she was a woman and a Jew. Her father encouraged her career in science, but her mother thought she should be a schoolteacher.

Cohn was recognized to be an exceptionally bright child and moved rapidly through the public school system, entering Hunter College, a free all-women's college in Manhattan, at age fifteen. She loved physics but majored in chemistry, since physics was not offered. Although the chairman of the Chemistry Department told her it was not ladylike for women to be chemists, she ignored him, and this would not be the first time she ignored such advice! On graduation from Hunter College in 1931, she enrolled in Columbia University where she studied with Harold Urey (Nobel Prize in Chemistry for the discovery of deuterium in 1934). She said that of all the great scientists she has known, Urey had the fastest mind, and she prepared herself carefully for a discussion with him, otherwise, she would be on point 2 when he was on point 5! She found Urey enthusiastic and inspiring and decided that she would like him as her research mentor. However, Columbia would neither give her a scholarship nor a teaching assistantship—only men were assigned the latter. To eke out her money for tuition, she lived at home and did babysitting jobs. She got as far as a master's degree, but then her money ran out and she took a job with the National Advisory Committee of Aeronautics (the precursor to NASA) at Langley Field in Virginia, where her project was to design an aircraft engine that ran on diesel fuel. She was the only woman among seventy men and was not promoted, because the director did not approve of women in research. Despite her unenviable position, the experience gave her a healthy respect for applied research.

After two years, she had saved enough money to return to Columbia. Urey tried to discourage her from working with him and said that he didn't mentor his students much and expected them to teach themselves. Once again, she ignored a professor's advice and told him that she wanted his mentorship notwithstanding. Urey finally gave in and sent her to his fellow professor, Isidor Rabi, to learn advanced physics. Rabi won the Nobel Prize in Physics in 1944 for devising a method for observing atomic spectra. Cohn began studies of isotope separation, starting with carbon and switching to oxygen. She failed to separate ^{12}C and ^{13}C, but this was due to equipment problems—the newly constructed mass spectrometer was malfunctioning. She got her PhD in physical chemistry in 1938 and continued working on the exchange of isotopic water ($H_2^{18}O$) with organic compounds (e.g., acetone). As a sideline, Urey wanted to determine the volume of water in the body and planned to use $H_2^{18}O$. Since Cohn was the smallest person in the laboratory and he wanted to conserve the isotope, she was asked to be the experimental subject, but this suggestion was not greeted with enthusiasm!

Research at Columbia University

Urey was supplying deuterium oxide (D_2O or heavy water) to two workers at Columbia College of Physicians and Surgeons. As noted earlier in this book, those individuals were Rudolf Schoenheimer, a Jew who had left Germany because of Nazi policies, and David Rittenberg, a former student of Urey's. Their work studying metabolism using stable, nonradioactive isotopes impressed Cohn because it yielded unequivocal information about metabolism in whole animals. The most important fact to emerge from their work was that compounds that were formerly thought to be static were actually turning over in the body.

As recounted in one of her recollections, in the spring of 1938, she was present at an encounter between two giants of physics, which she described as being of historic dimensions. While she was chatting with Urey at the annual meeting of the American Physical Society in Washington, E. O. Lawrence of the University of California, Berkeley approached them. Lawrence invented the cyclotron and would win the Nobel Prize the next year for his work on atomic structure. Urey introduced Cohn to him and then started uncharacteristically to taunt Lawrence by pointing out that his group had no radioactive isotopes suitable for biological work. Urey went on to say: "Hydrogen, we have deuterium, carbon, we have ^{13}C, you have ^{11}C with a half life of 20 minutes, nitrogen, we have ^{15}N, you have nothing and oxygen, we have ^{18}O and you have nothing." After this remarkable interchange, Lawrence told his group at Berkeley that they had to find a radioactive isotope with a long enough half-life to be useful to biologists. Shortly thereafter, ^{14}C was produced and transformed the study of metabolism forever!

Move to Du Vigneaud Laboratory

Following completion of her PhD, Cohn started looking for a job. It was the worst of the Great Depression, and Urey regretted that he was unable to find her a job. As he sadly put it, "Nobody wants you." At this time, most PhDs took jobs in industry, and the large companies would send recruiters to Columbia. Their notices would specify that applicants should be male and Christian, and she was neither. However, with Urey's help and that of Rittenberg and Schoenheimer, Cohn obtained a post-doctoral position at George Washington University with Vincent du Vigneaud, who would win the Nobel Prize in Chemistry in 1955 for his work on biologically important sulfur compounds and the synthesis of the hormone oxytocin. He was interested in the chemistry of insulin and had trained under John Jacob Abel at Johns Hopkins, where he realized that insulin was a protein, and his work had extended to other compounds containing sulfur. In du Vigneaud's laboratory, Cohn utilized isotopic tracers to follow sulfur amino acid metabolism and also studied transmethylation. Du Vigneaud's style of running a laboratory was very different from what she expected. Individual researchers would do only a limited part of a project, with du Vigneaud coordinating all the parts. Furthermore, if you suggested an experiment, he would make the decision as to whether it would be pursued. This style was quite different from what she experienced as a graduate student. Despite the adjustment she had to make, she spent nine years working with du Vigneaud. She developed great enthusiasm for the idea of applying isotopes to biological problems and never regretted her move from chemical physics to biology.

Cohn tells how she was dismayed during the first week in du Vigneaud's laboratory when the great man asked her to repair a piece of equipment, presumably because of her background in physical chemistry, and that he was disappointed when she said she couldn't do it. She was vindicated when a repairman for the company told him: "I wouldn't touch it with a ten-foot pole." A few months later, du Vigneaud installed an internal phone system, and the researchers were petrified that "Big Brother" could overhear their conversations. He asked Mildred to modify the system so that that would not be possible. She was appalled since she knew nothing about telephone circuits. She consulted a friend who had an electrical engineer look at it. When du Vigneaud learned of the cost of the modifications, he promptly dropped the project. Shortly after the laboratory moved to Cornell Medical College in 1938, Cohn built an electrophoretic apparatus (used to separate proteins and other compounds), which ran on 10,000 volts. As noted below, Cohn also built a mass spectrometer, which separates and measures compounds on the basis of their mass. In her work with du Vigneaud, this was used to follow the metabolism of components labeled with stable isotopes.

While in Washington, she married Henry Primakoff, a friend from her Columbia days and a theoretical physicist. Their marriage lasted until his death in July 1983. He provided great support to her, although it wasn't always helpful, as will be discussed below.

Studies of Amino Acid Metabolism

Much of the work in the du Vigneaud laboratory involved feeding amino acids and other compounds to rats and watching what happened. Methionine is an essential amino acid i.e. the body cannot make it, so it must be supplied in the diet. But there was a report that homocysteine could replace it in the diet. So homocysteine was labeled with deuterium and given to rats along with all the other amino acids (minus methionine) plus vitamins. However, the animals failed to grow, indicating that there was no conversion to methionine. On the other hand, when the same experiment was carried out by W. C. Rose in Urbana, Illinois, the animals thrived. Rose was renowned for his work on amino acid metabolism and had discovered the amino acid threonine. The difference in the findings of the two groups was that Rose used tiki tiki, a crude source of B vitamins, instead of pure B vitamins. Because the du Vigneaud animals showed fat infiltration in their livers, a sign of choline deficiency, he made the intellectual leap that choline was present in the tiki tiki mixture and was required for transmethylation—that is, the transfer of methyl groups to homocystine to make methionine. Cohn was stunned by this example of deductive reasoning.

At those times, biochemical reagents, including many amino acids, were not readily available, and, during June and July, all research stopped when the graduate students and postdoctoral fellows had to synthesize the ten amino acids not commercially available. Despite her lack of training in organic chemistry, Cohn had to synthesize deuteriomethyl-methionine for transmethylation studies. Luckily, a chemistry professor at Columbia allowed her to make deuterated methanol, but this had to be done in his laboratory since it required performing the reaction at 10,000 pounds of pressure, using a catalyst at 300°C and running the reaction for a month! The deuterated methanol was then used to make the labeled methionine, which was used to prove that methionine donated methyl groups to choline and creatine.

Work in the du Vigneaud laboratory required a lot of feeding of rats with isotopic compounds and special diets. Sometimes it took a year for a special compound to be synthesized. Mildred describes having to feed some rats one of these precious compounds on a weekend when no one was available to help her. As an emergency measure, she had to rely on her husband, a theoretical physicist, and instructed him how to hold the rat. She managed to insert the catheter through which the liquid diet would be administered, but failed to tell her husband that this often caused the animals to urinate and defecate, which this particular animal did! Surprised by this, her husband promptly dropped the rat, which took off with the catheter trailing. It took them over an hour to retrieve it, but the experiment was a success!

One of the most elegant tracer experiments with which Cohn was involved dealt with the question of the pathway of conversion of methionine to cysteine. The question was whether the sulfur or carbon atoms were used. The experiment required the synthesis of methionine appropriately labeled with ^{13}C and ^{35}S, and this took a year to prepare. However, the results showed unequivocally that the sulfur in the cysteine was derived from the sulfur in methionine, but not the carbon. When Cohn was carrying out her elegant tracer studies to elucidate the metabolism of sulfur-containing compounds and to study methyl transfer, only stable nonradioactive isotopes were available, and two of these required the use of a mass spectrometer. Du Vigneaud decided that his laboratory should have its own, and Mildred was given the task of constructing it. A scientist at McMaster University in Canada was the only person available to help, and he generously provided information on the components. Since it was wartime, Cohn had great difficulty getting components and had to improvise a lot, and the machine behaved somewhat erratically. When radioactive isotopes (^{14}C, ^{3}H, ^{32}P, ^{125}I, ^{35}S) became available, tracer technology for studying metabolism was greatly simplified. However, stable isotopes are still used in human studies because of the absence of radioactivity. Cohn's recollections are filled with amusing anecdotes. One involved du Vigneaud conducting a tour of the research laboratories with some medical students, trying to entice them into research. When he had described the mass spectrometer and its uses, he told the students that it had been built by Cohn, whereupon one of the students asked, "But can she cook?" Du Vigneaud responded, "I don't know, but she has two children."

Move to the Cori Laboratory

In 1946, Cohn went to the Cori laboratory where she was one of the first to use nuclear magnetic resonance (NMR) and electron paramagnetic resonance (EPR) to study enzymatic reactions. When she arrived in St. Louis, Gerty Cori gave her a warm welcome and took her under her wing. Gerty commented, "I understand you are more fortunate than I; you have a daughter and a son; I have only a son." In the Cori laboratory, Cohn was determined to use isotopes not merely as tracers of metabolism, but also for insight into the mechanisms of enzymatic reactions. After her experience with du Vigneaud, Cohn was also determined to work independently.

Carl Cori endorsed both of her objectives. One of the first projects was to use ^{32}P and ^{14}C to see if phosphorylase catalyzed an exchange of either inorganic phosphate or glucose with glucose-1-phosphate—it did not. On the other hand, experiments utilizing ^{32}P and ^{14}C supported the mechanism of the phosphoglucomutase proposed by Leloir (described in chapter 8). Working independently, Cohn used ^{18}O to determine which bond was cleaved when glucose-1-phosphate was subjected to chemical or enzymatic hydrolysis. In acid hydrolysis, the C-O bond was cleaved, whereas when acid or alkaline phosphatase was used, it was the P-O bond. When phosphorylase was used, the C-O bond was cleaved. This led to the general conclusion that phosphorylase enzymes were glucosyl-transferring, but the phosphatases were phosphoryl-transferring. A similar approach was used to explore the mechanisms of other enzymes that transfer phosphoryl groups. The experiments with ^{18}O necessitated her constructing yet another mass spectrometer.

In 1950, Cohn decided to spend some time with Paul Boyer at Harvard and to extend her work with ^{18}O to more complex systems such as oxidative phosphorylation. Her approach was to track the loss of ^{18}O from inorganic phosphate during oxidative phosphorylation in rat liver mitochondria. She found that during the oxidation of several substrates, there was a replacement of the ^{18}O in inorganic phosphate with ^{16}O, and this paralleled the phosphorylation that accompanied the oxidation. She deduced from this that the only source of ^{16}O that was large enough to produce the change was water, and therefore, it must be involved in the reaction. Further work on the phosphate-water exchange occurring during oxidative phosphorylation proved that it could not be the same as the substrate-level phosphorylation of glycolysis in which a C-O-P intermediate forms.

During her work on oxidative phosphorylation, Cohn decided to tackle a less intractable problem. This was the mechanism of kinases, i.e. enzymes that utilize ATP to phosphorylate protein substrates. These were known to require divalent metal ions, usually Mg^{2+}. Cohn reasoned that she could determine if the site of bond cleavage was at the α- or β-P of ATP by determining the structure of the metal chelate involved in the reaction. Furthermore, she felt that paramagnetic Mn could substitute for Mg, but she found that EPR could not detect the formation of a complex, illustrating that brilliant experiments do not always work out in practice. She was later successful, using NMR.

In 1955, Cohn spent a sabbatical in the laboratory of Hans Krebs at Oxford, where she got the idea that she could elucidate the structure of metal-ATP-enzyme complexes using ^1H and ^{31}P-NMR. When she got back to St. Louis, she found that the NMR spectrometer was limited to ^1H. She wrote to Varian, the company in Palo Alto that made the machines, asking for permission to use one of theirs. They invited her immediately, however with her three small children, it was impossible to do so. Two years later, her husband was invited to Stanford as a visiting professor, and the whole family spent the summer in Palo Alto. When she contacted Varian again, she got a cool reception because the machines were all in use. Nevertheless, they allowed her access, but only for two separate days. The results indicated that the approach was feasible; the adenine and ribose protons were resolved as were the α-, β- and γ-P of ATP. Furthermore, Mg and Mn produced large changes in the spectra. Cohn said that one of the most exciting moments in her scientific career was when she saw the first three peaks (P atoms) of ATP using NMR.

Work with Britton Chance at the Johnson Foundation

In 1960, Cohn's husband accepted a position as Donner Professor of Physics at the University of Pennsylvania, and she joined the Johnson Foundation under the direction of Britton Chance. Chance was a remarkable person, with a love of the sea that manifested itself in a love of sailing and the invention, as a teenager, of an automatic steering device for ships. His intense competitive spirit and passion for sailing earned him a spot on the 1952 Olympic yacht team, where he won a gold medal. As a graduate student at the University of Pennsylvania, Chance became interested in stop-flow methods to study very rapid reactions and wanted to go to Cambridge University to work with Glenn Millikan, who had developed a novel apparatus for doing this. Millikan was the son of Robert Millikan, who won the Nobel Prize in Physics in 1923 for measuring the charge on the electron using the famous oil droplet method. Glenn Millikan would later be appointed to the chair of physiology at Vanderbilt Medical School, but soon thereafter, he lost his life in a rock climbing accident in East Tennessee. As an interesting aside, he was married to the daughter of George Mallory, the English mountaineer who lost his life in one of the earliest attempts to summit Mt. Everest. I mention this because one of Millikan's successors in the chair of physiology was Charles Park, who went to Vanderbilt after spending five years in the Cori laboratory.

Britton Chance obtained a PhD from Cambridge University in 1942, but his time in England was interrupted by World War II when he returned to work in the radiation laboratory at the Massachusetts Institute of Technology, where he was involved in the development and enhancement of radar. When Chance returned to the University of Pennsylvania, he continued working on the stop-flow equipment. He used it to prove the mechanism of many enzyme reactions, for example, the existence of the enzyme-substrate complex. This was shown by demonstrating the existence of an intermediate compound in the reaction by which peroxidase catalyzes the breakdown of hydrogen peroxide and the determination of the rates of formation and breakdown of the compound. Furthermore, Chance used the kinetic data to prove the validity of the Michaelis-Menten equation.

At the Johnson Foundation, Chance was interested in energy transduction, oxidative phosphorylation, and ATP utilizing reactions. He was supportive when Cohn decided to explore the structural changes in ATP and ADP induced by divalent metal ions using NMR. She knew that Mg^{2+} and other such ions were involved in many reactions involving the nucleotides, but didn't know why. She first showed that chemical shifts of the NMR peaks of the phosphorus nuclei of ATP and ADP could be measured, and that the same techniques could be used to observe the changes in the chemical shifts of the phosphorus NMR spectra in the presence of metal ions. These provided direct evidence of the nature of the complexes between the nucleotides and the metal ions. Likewise, the changes in the chemical shifts of the proton magnetic spectra could determine the interaction of the metal ions with the adenine ring. Her paper describing this work was rejected by the *Journal of the American Chemical Society*, but accepted by the *Journal of Biological Chemistry*, where it became a citation classic.

During her time in the Johnson Laboratory, Cohn attempted to exploit every feasible aspect of magnetic resonance to elucidate the structure and function of enzymes, with a focus on kinase enzymes. The results yielded some surprises, including that the free energy change on the surface of kinase enzymes is close to zero and that the significant changes are involved in substrate binding and/or dissociation of products. This remarkable finding appeared to be general for many enzyme reactions. Other studies of the role of Mn^{2+} (substituting for Mg^{2+}) in the reactions of kinases revealed two classes of enzyme: those in which the metal ion was bound only to the nucleotide or was bound directly to the enzyme. In later work with Marianne Grunberg-Manago, Cohn used NMR to study the mechanism of the synthetase enzyme that utilizes ATP to synthesize methionyl tRNA. The significance of this reaction in protein synthesis is described in the chapter devoted to Severo Ochoa.

In the early stages of her career, Cohn did not receive the respect that her brilliant mind deserved. She was a research associate for twenty-one years and only achieved faculty rank when she was appointed an associate professor after twelve years in the Cori department. On moving to the Johnson Foundation, she was promoted to full professor after one year and awarded a Career Investigatorship of the American Heart Association (Figure 15). She retired as Benjamin Rush Professor Emerita of biochemistry and biophysics in 1982. Until the end of her days, she remained involved in science and in the life of the University of Pennsylvania, and described herself as being out of scientific research, but not out of science.

Despite her early exclusion from the academic ladder, she recognized that there were benefits. For example, she could spend time with her three children, especially if they were sick, and lengthy vacations were not a problem. She could tackle technically difficult, long-range problems without needing to compete or feel pressured to publish. In short, she described her experiences as fun, especially when an entirely unexpected phenomenon was discovered serendipitously or when the results could be applied to a medical problem. She noted that her life in science provided a stimulating environment with the opportunity to interact with first-class minds.

In her numerous biographical reminiscences, she displayed a great memory and a sense of humor. She described her earliest memory when, as a child, she was riding in a horse and carriage when the horse sat down. She was terrified because she had never seen a horse sit down! One of her students noted that in her professional life she was Mildred Cohn, but in her personal life she was Mildred Primakoff, and that when she edited papers, she signed her comments MCP (Mildred Cohn Primakoff), which at that time meant "male chauvinist pig"! When she was inducted into the National Women's Hall of Fame, she noted that Hilary Clinton and Oprah Winfrey were also members and added: "I decided this could be a good place for me."

Cohn's contributions were recognized by numerous awards, chiefly the National Medal of Science in 1982 and the Garvan Medal of the American Chemical Society. She was also elected to the National Academy of Sciences, the American Academy of Arts and Sciences, and the American Philosophical Society. She served as president of the American Society of Biological Chemists (now the American Society for Biochemistry and Molecular Biology). She was also one of the first women to be appointed to the editorial board of the *Journal of Biological Chemistry*. Cohn died in Philadelphia on October 12, 2009, at the age of ninety-six.

References

Chance, B. 1943. "The Kinetics of the Enzyme-Substrate Compound of Peroxidase." *J. Biol. Chem.* 151: 553–77.
Cohn, M. 1949. "Mechanism of Cleavage of Glucose-1-Phosphate." *J. Biol. Chem.* 180: 771–81.
Cohn, M. 1953. "A Study of Oxidative Phosphorylation with O^{18}-Labeled Inorganic Phosphate." *J. Biol. Chem.* 201: 735–50.
Cohn, M. 1992. "Atomic and Nuclear Probes of Enzyme Systems." *Ann. Rev. Biophys. Biomol. Struct.* 21: 1–26.
Cohn, M. 1995. "Some Early Tracer Experiments with Stable Isotopes." *Protein Science* 4: 2444–47.
Cohn, M. 2002. "Molecular Biophysics. Postdoctoral Years." *J. Biol. Chem.* 277: 10747–52.
Cohn, M., and G. T. Cori. 1948. "On the Mechanism of Action of Muscle and Potato Phosphorylase." *J. Biol. Chem.* 175: 89–93.
Cohn, M., A. Danchin, and M. Grunberg-Managò. 1969. "Proton Magnetic Relaxation Studies of Manganous Complexes of Transfer RNA and Related Compounds." *J. Mol. Biol.* 39: 199–217.
Cohn, M., and T. R. Hughes. 1962. "Nuclear Magnetic Spectra of Adenosine Di- and Tri-Phosphate. II. Effect of Complexing with Divalent Metal Ions." *J. Biol. Chem.* 237: 176–81.
du Vigneaud, V., G. W. Kilmer, J. R. Rachele, and M. Cohn. 1944. "On the Mechanism of Conversion in vivo of Methionine to Cystine." *J. Biol. Chem.* 155: 645–51.
du Vigneaud, V., C. Ressler, and S. Trippett. 1953. "The Sequence of Amino Acids in Oxytocin, With a Proposal for the Structure of Oxytocin." *J. Biol. Chem.* 205: 949–57.
Nageswara Rao, B. D., and M. Cohn. 1977. "Asymmetric Binding of the Inhibitor Di(adenosine-5') pentaphosphate to Adenylate Kinase." *Proc. Natl. Acad. Sci. U.S.A.* 74: 5355–57.
Rittenberg, D., and R. Schoenheimer. 1937. "Deuterium as an Indicator in the Study of Intermediary Metabolism: IX. Further Studies on the Biological Uptake of Deuterium into Organic Substances, With Special Relevance to Fat and Cholesterol Formation." *J. Biol. Chem.* 121: 235–53.
Schoenheimer, R., and D. Rittenberg. 1935. "Deuterium as an Indicator in the Study of Intermediary Metabolism." *J. Biol. Chem.* 111: 163–68.

13

Christian de Duve

Belgian with Savoir Faire

Early Education and Introduction to Research

Christian de Duve spent only four months in the Cori laboratory, working principally with Earl Sutherland in studies on glucagon. He was born on October 2, 1917, in Thames Ditton, a town near London. His parents, of Belgian-German extraction, were Belgian nationals who had taken refuge in England during World War I. They returned to Belgium in 1920, and de Duve grew up in Antwerp, where the education was a mixture of Flemish and French. He attended a Jesuit school and enjoyed learning Latin, Greek, philosophy, and mathematics, but he disliked history and geography. Physics, chemistry, and biology were poorly taught because the teachers had little interest in them. As a reflection of the bilingual nature of education in Belgium at that time, Latin was taught in French, Greek in Flemish, mathematics in French, and history in Flemish! De Duve excelled in school and loved to solve problems and answer questions, but this did not make him popular.

When it was time to choose a career, none of the humanities nor law or business appealed to him, but he was attracted to the romantic image of a physician presented in a movie he had seen. He decided on medical studies and entered the Catholic University of Louvain in 1934. Due to his family background and travel through Europe, he became proficient in four languages. As a second-year medical student in Louvain, he followed tradition where the "good" students spent their free time working in the laboratories of one of the professors. He chose the physiological laboratory of Joseph Bouckaert, who had trained with A. V. Hill in the energetics of muscle contraction. Bouckaert was a thinker rather than a bench scientist and relied on an assistant, Pierre-Paul De Nayer, to run the practical classes and supervise much of the research work. Bouckaert had an uncompromising commitment to pure basic research, which he saw as an activity entirely devoted to the pursuit of truth—free of bias and devoid of any lucrative or even utilitarian purpose. Although the university had strong ties to the Catholic Church, Bouckaert had a strictly mechanistic conception of living processes—a strong conviction that all life phenomena had to be explained in strictly physical and chemical terms without involving any sort of "vital force."

Bouckaert had no pet topic and felt that the research performed in his laboratory should cover every component of his physiology course, which he taught single-handed in both French and Flemish! His encyclopedic knowledge and rigorous analytical mind had a profound influence on de Duve. Bouckaert had groups working on muscle contraction, basal metabolism, kidney function, gastric secretion, cardiac activity, neurobiology, and insulin action. De Duve chose to join the insulin group, in part because he was attracted by the surgery (liver removal) involved in the project. Although insulin had been discovered in 1922, the mechanism by which it lowered blood glucose was still highly controversial. One school of thought believed it was due to an effect on the liver to inhibit glucose output or stimulate glucose uptake, whereas the other believed it was due to increased glucose uptake by muscle. At the time de Duve joined the fray, there was general agreement about the peripheral (muscle) effect, but the role of the liver was hotly debated.

Bouckaert recognized that the in vivo experiments in which insulin was injected were complicated by the release of epinephrine due to the hypoglycemia. To avoid this artifact, he devised a procedure known as the "compensation" technique, in which glucose was infused to keep the blood glucose constant and thus avoid the effects of hypoglycemia. The amount of infused glucose needed to do this was a measure of the amount of glucose consumed by the animal. The experiments were repeated using animals whose livers or abdominal viscera had been removed. At this stage, de Duve became involved in the experiments, and the most striking finding was that when insulin was injected into the animals without livers, they needed much less infused glucose to keep their blood glucose levels normal. In other words, the experiments indicated an inhibitory effect of insulin on glucose output by the liver.

A pleasant daily event in the Bouckaert laboratory was the "goûter," a continental version of an English tea, which was served in the library. It consisted of large slices of bread spread with margarine and a syrupy concoction called "sirop de Liége," washed down with a pallid chicory extract euphemistically called coffee. Despite the unattractive nature of the repast, the goûter was the center for the exchange of ideas. New results were presented for general discussion so that everyone knew of the research being done by others, and interesting comments and criticisms were offered. When the scientific discussions were over, there often followed a conversation on topics of general interest. This was led by Bouckaert, who had a wide range of interests and knowledge, drawn largely from the *Encyclopedia Britannica*, of which he was an avid reader.

De Duve enjoyed both the physical aspects of research, such as the exercise of manual skill that was involved in the performance of delicate and precise surgery, and the intellectual aspects, for example, solving puzzles and discovering something that nobody had known or seen. Being trained in rigorous logic by the Jesuits, he enjoyed the rules by which the scientific method was applied. In one of his autobiographies, he describes the three cardinal rules for the design of experiments. First, think beforehand so that you know how to conduct the experiment, but not to the extent that you are not alert to the unexpected or abnormal. Second, do not set out to prove a hypothesis, but rather to test it and try to disprove it—ideas coming from the philosopher Karl Popper and the physiologist Claude Bernard. Third, do not repeat what has already been done by spending time reviewing the literature. In the next phase, when the data have to be analyzed and interpreted, this challenging exercise requires a mixture of rigor and imagination tempered by an open mind. The final phase involves writing up the work for publication, and here de Duve recognizes the value of his training in the humanities with emphasis on proper terminology, analytical precision, and syntactic rigor, but admits to a temptation to embellish his style with ponderous adjectives and tedious asides.

He found research gripping and soon gave up his interest in becoming a physician and decided instead to devote himself to a life in research, although he recognized that obtaining an academic position in Belgium was extremely difficult at that time—he jocularly said that you had to be the son of a professor or a nephew of a bishop to obtain such a position. But this did not deter him. He felt that his fluency in several languages—especially French and Flemish—would be a great asset. Surprisingly, Bouckaert did not think that de Duve had any chance of securing an academic position in Louvain. However, de Duve had been incurably infected with the virus of research and resolved to pursue a career in academic medicine.

Interest in Insulin Action on Liver

De Duve graduated from medical school in 1941, after World War II had broken out; Belgium was occupied by the Nazis after the invasion of May 10, 1940. The citizens of Louvain were outraged when Stuka dive bombers attacked the city, especially the historic library. This beautiful building had been destroyed by the Germans in World War I and rebuilt with funds collected from students at American schools, colleges, and universities. De Duve had a brief stint in the Belgian Army and was captured, but managed to escape from a prisoner's column by simply stepping on the running board of a passing car! Despite the difficulties imposed by the war conditions, he returned to Louvain and continued to pursue his goal of elucidating how insulin affected the liver. He decided that he needed to go back to school and learn more chemistry and supported himself by working as a clinical assistant at a cancer institute. Because of the difficulties associated with experimental work, he decided to write up his work and wrote a book titled "*Glucose, Insuline et Diabète*," with 400 pages and 1,200 references. He eventually condensed this into a dissertation, which he presented for the degree of Agrégé de l'Enseignement Supérieur, which is roughly equivalent to a PhD and is required for a professorial position. He also wrote a review on the laboratory's work on insulin, which was written in English and published in *Physiological Reviews*.

In 1943, de Duve married the daughter of a doctor who practiced in a small city in Belgium; as of this writing, they are still happily married. His wife's family was interesting because one of her cousins married Aldous Huxley, the author of *Brave New World* and other famous novels.

When de Duve was writing his articles, he developed the (incorrect) notion that the rate of glucose utilization by the tissues was directly proportional to the blood glucose concentration and was independent of any hormonal influences. This idea was supported by the work of the American investigator Samuel Soskin, who championed the idea that there was a "hepatic threshold" for glucose, that is, a blood glucose concentration at which glucose production by the liver was exactly balanced by glucose utilization. This ran counter to the ideas of the Coris, who believed that insulin promoted the transfer of glucose from the liver to the periphery. In point of fact, this effect of "insulin" was due to its contamination with glucagon, a hormone that would later occupy much of de Duve's research.

Toward the end of the war, allied bombers planned to attack the railway marshalling yards of Louvain, but the wind blew in the wrong direction and the city itself was bombed with tragic consequences. After the wounded had been taken care of, de Duve walked through the rubble-strewn streets. He was saddened by the devastation but uplifted by the fact that the Allies would soon liberate the city. As he walked, his thoughts went back to the so-called liver paradox, namely, that despite their demonstration that insulin decreased glucose output from the liver in vivo, some investigators had shown that insulin increased glucose release from liver slices. De Duve's thoughts became focused on glucagon as the reason for the difference since he knew that this was an impurity in certain preparations of insulin. There were great discrepancies between the findings of various groups with respect to the effects of insulin on the liver, and de Duve deduced that these differences were due to the types of insulin used in the experiments. The insulin was obtained from Eli Lilly, Allen and Hanburys (a British company), or the Danish company Novo, and de Duve speculated that the Eli Lilly preparation was contaminated with glucagon. However, due to the war, he was unable to obtain the Lilly insulin to test his idea. When the Allied forces entered Louvain, de Duve was able to obtain some of this from an American doctor. With the assistance of a medical student, Henri-Géry Hers, de Duve injected a rabbit and found an increase in the blue color when he tested the blood using the classic Folin-Benedict reaction for glucose, indicating hyperglycemia. When he tested the Novo insulin, he observed no such reaction. In his autobiography, he states that he felt like the French astronomer Le Verrier, who had discovered the planet Neptune in the location that he predicted. The difference between the Lilly and Novo insulins was later confirmed using Bouckaert's compensatory glucose infusion technique.

Training with Hugo Theorell in Stockholm

After the liberation of Belgium, de Duve devoted his time to training in biochemistry. He obtained a master's degree in chemistry—his project was the purification of penicillin—and then spent eighteen months in the laboratory of Hugo Theorell in Stockholm, working on the crystallization of human myoglobin, which entailed the gruesome task of grinding up human hearts obtained from an autopsy room. Theorell was a pupil of Otto Warburg and had isolated, purified, and crystallized no fewer than six key enzymes and used them for precise kinetic studies involving different cofactors. He would win the Nobel Prize in 1955 for his work on these enzymes. De Duve was unsure why Theorell accepted him since he was totally untrained in biochemistry. Perhaps it was that he brought his own funding and that he spoke French, which Theorell liked immensely and used in lunchtime conversations with him. De Duve learned a lot from Theorell's head technician, who eventually rose to earn a PhD and a professorship. Nothing was disposable—a far cry from present-day laboratory procedures—and pipettes had to be laboriously cleaned by immersing them in hot chromic acid, followed by washing in tap water and distilled water, and finally rinsing and drying in alcohol and ether. The prize instrument of the laboratory was a huge spectrophotometer in front of which Theorell sat analyzing spectra. Theorell was crippled by poliomyelitis and needed crutches, and sometimes he walked on his hands!

Sune Bergstrom and Bengt Samuelsson spent time in the Theorell laboratory while de Duve was there. Together with John Vane, they would win the Nobel Prize in 1972 for their work on prostaglandins. The laboratory attracted other promising scientists including Christian Anfinsen, who won the Nobel Prize for chemistry in 1972 with William Stein and Stanford Moore for work on the sequence and structure of RNA enzymes. A colorful visitor, already described in chapter 12, was Britton Chance. He introduced his stop-flow technique for analyzing fast enzymatic reactions. Chance was a former Olympian and took Theorell sailing around the islands of the Stockholm Archipelago. Despite his disability, Theorell was an enthusiastic sailor. Other visitors were John Buchanan, an expert on purine biosynthesis, and Ralph Holman, who specialized in lipids.

Move to the Cori Laboratory

De Duve's dream was to return to his love affair with insulin, and he resolved to go to the Cori laboratory, even though they were proponents of a theory that he believed to be false, namely, that insulin stimulated glycogen breakdown in the liver. He had been appointed to a lectureship in physiological chemistry at his alma mater in Louvain, but this permitted him to spend a few months with the Coris. He wrote to Carl Cori, but Cori's response was coldly negative, indicating that he never accepted anyone at his laboratory for less than a year. He added: "With you, there is the added difficulty that we do not see eye to eye with respect to the mechanism of action of insulin." All was not lost when a second letter came from Cori, stating: "You may be interested to hear that Dr. Earl Sutherland, in my laboratory, has just discovered that the glycogenolytic effect of insulin on liver slices is due to an impurity. I believe you have also done some work on this subject." Cori then invited de Duve to come to St. Louis to collaborate with Sutherland. Cori had read a paper that described the stimulation of glycogen breakdown in liver slices incubated with insulin and he had asked Sutherland to investigate this. As described chapter 7, Sutherland found that the glycogenolytic effect was not due to insulin itself since it persisted when insulin had been inactivated. De Duve had difficulty obtaining support from Belgian sources for his stay with Cori, but Theorell solved the problem by obtaining a fellowship for him from the Rockefeller Foundation.

De Duve spent only four months in the Cori laboratory in 1947. He described his experience as unforgettable, although he didn't see much of the Coris. In part, this was due to their being awarded the Nobel Prize, which had happened one month after de Duve's arrival. The announcement from Stockholm created much excitement, and it was helpful to the Coris that de Duve and his wife had attended the ceremonies the year before at the invitation of Theorell. Thus de Duve's wife could advise Gerty, who paid little attention to her appearance, on what to wear. De Duve remembered being quizzed by Theorell, a member of the Nobel Committee on Medicine, about the achievements of the Coris. The Committee was impressed with their report that phosphorylase could synthesize glycogen from glucose-1-phosphate. Although their accomplishments were many and significant, this particular finding was an artifact from a physiological point of view, and in vivo phosphorylase acts to break down glycogen!

Research with Earl Sutherland

In his autobiography, de Duve describes the Cori laboratory as a beehive in which small individual groups worked in separate laboratories on distinct problems. He worked almost entirely with Sutherland and notes that their temperaments were very different. He describes Sutherland as outwardly jovial and easygoing, with a deceptively casual and self-deprecating attitude that rarely betrayed the passion that drove him. Only when he was opposed and convinced that he was right did he become obstinately unyielding. Sutherland's approach to research was essentially pragmatic and intuitive, seeming to rely on instinct and flair rather than on rational reasoning. He had a keen mind but a baffling, circumlocutory way of expressing himself, often voicing only the final conclusion of a long internal monologue, which had to be guessed at, but when divined, was very pertinent. When de Duve and Sutherland set out independently to discover the mechanism of the effect of glucagon on glycogenolysis, Sutherland unerringly chose phosphorylase as the target, whereas de Duve discarded it since it was thought to synthesize glycogen.

Despite their personality differences, de Duve and Sutherland made a great team and became close friends. They shared a burning interest in their project and an almost untiring capacity for work, often laboring until late at night and ending up at an insalubrious diner called "The Pig's Ear." In four hectic months, they showed that only the pancreas, stomach, and certain other parts of the gastrointestinal tract made glucagon and that it was produced in the pancreatic islets by cells different from the beta cells that made insulin. Despite the cumbersome name, Sutherland persisted in calling glucagon the hyperglycemic-glycogenolytic factor. For some strange reason, he took a great dislike to the name glucagon. On this point, it should be noted that in 1923, soon after the discovery of insulin, John Murlin at the University of Rochester reported that he had found a hyperglycemic factor in pancreatic extracts that he named glucagon. It is unclear when Sutherland and Cori became aware of this report.

The Coris refused to be coauthors of the Sutherland-de Duve paper because they believed that they had not really participated. When de Duve suggested that the paper be split in two so that he and Sutherland could be first authors, Sutherland adamantly refused to do this.

De Duve notes that his stay in St. Louis brought other dividends besides working in the foremost biochemical laboratory at that time. For example, the Eli Lilly Company, which had taken over the purification and crystallization of glucagon, began to notice de Duve and his work. The company also provided him with support and invited him to their annual conferences on insulin, held in Indianapolis. Before returning to Belgium, de Duve visited some prominent scientists in the fields of insulin and metabolism. These included Charles Best in Toronto, Rachmiel Levine in Chicago, and Baird Hastings in Boston. He also called on Fritz Lipmann and paid a visit to the famous Rockefeller Institute in New York, where he met his fellow Belgian Albert Claude, who introduced him to the young George Palade. Little did de Duve realize that all three would jointly receive the Nobel Prize in 1974 (Figure 16).

Return to Belgium and Discovery of Lysosomes and Peroxisomes

When de Duve returned to Louvain in 1947 to take over the teaching of physiological chemistry in the medical school, the Lilly Company provided him with the funds to set up a laboratory. This was helpful because the rector of the university, who was a bishop, said there was no money available. De Duve's major project involved the enzyme glucose-6-phosphatase, which is critical for the ability of the liver to release glucose as a result of glycogen breakdown. Its absence from muscle was the reason glycogenolysis in this tissue resulted in the release of lactate, but not glucose. Key members of de Duve's team were Henri-Géry Hers, Jacques Berthet and his wife Lucie, and a medical student, Henri Beaufay. Together, they set out to purify glucose-6-phosphatase using the rather primitive techniques available at that time, and their results indicated that the enzyme was attached to some intracellular structure. They then set about identifying this structure using the technique of differential centrifugation (centrifuging tissue homogenates at different speeds and times to separate intracellular organelles). This had been developed by de Duve's fellow countryman, Albert Claude, and provided information on the distribution of enzymes in the organelles, and also their structures as revealed by electron microscopy. In this, he was helped by Keith Porter, a premier electron microscopist, and George Palade, a young researcher from Romania.

At the end of 1949, de Duve and his associates did their first fractionation and found that glucose-6-phosphatase activity accompanied a fraction, which Claude called microsomes. Further work showed that the enzyme was firmly attached to a lipoprotein structure that is now recognized as the endoplasmic reticulum. Along with their studies of glucose-6-phosphatase, they monitored another nonspecific enzyme termed acid phosphatase because of its potential to interfere with their results. This enzyme showed an abnormally low activity when first assayed in liver homogenates, but after storage in a refrigerator for several days, its activity increased markedly. The way this was discovered was that Berthet and his future wife went to de Duve's office at 10:00 p.m., upset because the activity of the enzyme was much lower than expected. So de Duve told them to put the fractions in the refrigerator and assay them the next week. When they re-assayed the fractions, there was a large increase in activity. When they subjected the fraction to a variety of procedures to disrupt the particles, the enzyme was released, and they reasoned that its latency must be due to its being enclosed within vesicles that prevented the substrate from getting in or the enzyme from getting out. However, the situation was more complex, because the distribution of the enzyme in different subcellular fractions was unclear. Some of it was recovered in the mitochondrial fraction and some in the microsomal fraction. Eventually, they realized that acid phosphatase was associated with a new subcellular structure, which was subsequently called the lysosome. How this came about was somewhat serendipitous, as described below.

A young American named Berton Pressman joined the group from Madison, Wisconsin, where he had worked with Henry Lardy. Pressman was rather brash and not well-mannered, and believed that all their laboratory procedures were wrong. He studied a proteolytic enzyme, cathepsin D, which also demonstrated latency. The fractionation technique was now modified from that used by Claude to that used by two of his coworkers (Walter Schneider and George Hogeboom) in which the saline was replaced by isotonic sucrose and a more gentle homogenization procedure was used. The point was to preserve the cell organelles in a more intact state. When de Duve used this procedure, he was surprised to find that two-thirds of the acid phosphatase activity was associated with the mitochondria and one-third with the microsomes. He suggested three explanations for this strange result, one of which was that they were dealing with a new organelle. They explored this possibility by assaying the fractions obtained by centrifugation for acid phosphatase and enzyme markers for known organelles (cytochrome oxidase for mitochondria, glucose-6-phosphatase for microsomes). They also started varying the centrifugation protocol. Here, serendipity intervened when a centrifuge being operated by a medical student broke down, and a lower power machine had to be used, yielding no enzyme. This indicated that the particles containing the enzyme required sedimentation at a centrifugal force midway between that for mitochondria and microsomes. The particles were called lysosomes, which is Greek for digestive bodies, but de Duve somewhat regretted this because of the possible confusion with enzyme lysozyme. Interestingly, lysosomes had been discovered earlier by the Russian zoologist Ilie Metchnikoff, who had examined feeding in protozoa and had found that these digest their food into vacuoles (i.e., lysosomes). For his discovery of white blood cells, Metchnikoff shared the 1908 Nobel Prize with Paul Ehrlich, who originated the idea of the killing of bacteria with chemicals. Two years later, Ehrlich would achieve popular fame with his discovery of salvarsan, "the magic bullet," as a cure for syphilis.

De Duve and his associates searched for additional enzymes associated with lysosomes and found five others, which were all acid hydrolases. The presence of these enzymes suggested a digestive function for lysosomes. Another enzyme was initially thought to be associated with lysosomes, but it did not demonstrate latency. As described below, this enzyme, urate oxidase, was eventually traced to another particle. The morphology of lysosomes was established in collaboration with the American microscopist Alex Novikoff, and interest in them spread rapidly. For the discovery of these organelles, de Duve was awarded the Nobel Prize in 1974, together with Albert Claude and George Palade. Lysosomes achieved great importance in medicine when a variety of diseases were traced to lysosomal enzyme deficiencies, the first being Type II glycogen storage disease (Pompe's disease), which was shown by Henri-Géry Hers to be caused by a lack of an acid α-glucosidase. Lysosomes are now recognized as the principal sites of intracellular digestion, and they contain more than forty types of hydrolytic enzymes, including proteases, nucleases, glycosidases, lipases, phosphatases, and sulfatases. Materials are delivered to lysosomes by multiple pathways, including uptake from the extracellular fluid by endocytosis or phagocytosis and by the process of autophagy, where phagosomes carry obsolete cell constituents and fuse with lysosomes. In the normal situation, materials entering the lysosomes by endocytosis or phagocytosis are broken down by the lysosomal enzymes, and the products are released. But if an enzyme is missing due to a genetic defect, materials such as glycogen, glucocerebrosides, glycolipids, and sphingomyelin accumulate in the lysosomes, resulting in storage diseases, which are frequently devastating.

After completing his work on lysosomes, de Duve returned to the study of urate oxidase, which was different from the lysosomal enzymes in terms of its lack of latency and hydrolytic activity. He found some other enzymes that behaved like it, including D-amino oxidase and catalase. Hydrogen peroxide was produced by the action of urate oxidase and D-amino oxidase, but catalase used hydrogen peroxide to oxidize a variety of compounds. Using similar approaches to those for lysosomes, but employing density gradient centrifugation (where particles are centrifuged through a fluid with a density gradient in which they end up in a layer equal to their density), de Duve and his colleagues identified a new particle, which they called the peroxisome. Later work showed that these are present in all eukaryotic cells and that they are involved in fatty acid oxidation and amino acid metabolism. Some diseases have been ascribed to their deficiency.

Appointment at the Rockefeller Institute and Establishment of the International Institute of Cellular and Molecular Pathology

In 1960, de Duve was beginning to tire of his administrative duties and developed the idea that he should join the Rockefeller Institute in New York. He told George Palade about this on a visit to the Institute. Palade looked at him quizzically, but de Duve said he meant it. A few months later, Detlev Bronk, the president of Rockefeller and previously president of the National Academy of Sciences, flew to Brussels and offered de Duve a job and a laboratory. De Duve went to see the rector of his university, who suddenly realized that he did not want to lose him. The rector offered de Duve a part-time appointment where he wouldn't have to teach or conduct exams. The rector invited Bronk to dinner and provided a repast that only a bishop could. The two reached a compromise by which, for five years, de Duve would split his time between New York and Louvain, and after that would be permanently based in New York. This arrangement became permanent in 1962, but his family remained in Belgium. The situation was productive from a scientific viewpoint, and many of de Duve's Belgian associates spent time with him in New York. In October 1974, while he was in New York, a member of the administration of Rockefeller University (the name was changed so that graduate students could work there) told him de Duve he had won the Nobel Prize together with his Rockefeller associates, Claude and Palade.

In 1968, the Catholic University of Louvain was divided into two separate universities. This was the result of the increasing divide between Flemish speakers and French speakers. Before the division, a single faculty could have both Flemish and French speakers. The split resulted in the building of the Université Catholique de Louvain (UCL) on a brand new campus in a French-speaking area closer to Brussels

(Louvain-la-Neuve). Because of the need for clinical training, the medical school was located on a different site on the outskirts of Brussels, and a new hospital was built. De Duve decried the separation of the medical faculty from the rest of the faculty in Louvain-la-Neuve. To strengthen the new medical school, he conceived the idea of an international, multidisciplinary research institute that would be associated with it. This was originally named the International Institute of Cellular and Molecular Pathology (ICP), but later received the less cumbersome title of the de Duve Institute. De Duve founded it on the basis of three principles: priority to basic research and freedom of the investigators, special attention to medical benefits resulting from basic discoveries, and multidisciplinary collaboration within a critical mass of competency. He began with four groups with backgrounds in cell and molecular biology, but the Institute expanded to include several laboratories of the medical faculty and collaboration with other laboratories on the campus. The building, a large multistory structure adjacent to the medical school and hospital, was opened in April 1975 by Prince Albert of Belgium. It was greatly strengthened when the Ludwig Institute for Cancer Research, founded by the American entrepreneur Daniel Ludwig, established a large institute within the ICP.

During the development of the ICP, de Duve still shared his time between Brussels and New York, but his involvement in research gradually came to an end. He became an emeritus professor at the UCL in 1985 and at the Rockefeller University in 1988. In 1991, he gave up his duties as president of the ICP, which had grown to 270 investigators. His interests have recently been focused on the origin and evolution of life, and he has written three books on the subject. In addition to the Nobel Prize, he won the Francqui Prize, Belgium's most prestigious prize in the field of science

References

Beaufay, H., and C. de Duve. 1954. "The Hexose Phosphatase System. VI. Attempted Fractionation of Microsomes Containing Glucose-6-Phosphatase." *Bull. Soc. Chim. Biol.* 36: 1551–68.
Bouckaert, J., and C. de Duve, C. 1947. "The Action of Insulin." *Physiol. Rev.* 27: 39–71.
Bouckaert, J. P., P. P. De Nayer, and R. Krekels. 1929. "Équilibre Glucose-Insuline." *Arch. Int. Physiol.* 31: 180–93.
De Duve, C. 1945. *Insuline et Diabète*. Paris: Masson et Cie.
De Duve, C. 1983. "Lysosmes Revisited." *Eur. J. Biochem.* 137: 391–97.
De Duve, C. 1994. "Born Again Glucagon." *FASEB. J.* 8: 979–81.
De Duve, C. 1996. "The Peroxisome in Retrospect." *Ann. N.Y. Acad. Sci.* 804: 1–10.
De Duve, C. 1997. "International Institute of Cellular and Molecular Pathology." *Molecular Medicine* 3: 87–89.
De Duve C. 2004. "My Love Affair with Insulin." *J. Biol. Chem.* 279: 21679–88.
De Duve, C. 2005. "The Lysosome Turns Fifty." *Nature Cell Biology* 7: 847–49.
De Duve, C. 2006. "How I Became a Biochemist." *Life* 58: 614–18.
De Duve, C., and J. P. Bouckaert. 1944. "Nouvelles Recherches Concernant l'Action de l'Insuline. I. Influence de la Hauteur de Glycémie sur l'Action de l'Insuline Chez l'Animal Normal." *Arch. Int. Pharmacodyn. Thér.* 69: 486–501.

De Duve, C., P. P. De Nayer, J. De Keyser, and J. P. Bouckaert. 1945. "The Action of Insulin. III. Effect of Different Conditions on the Hepatic Action of Insulin." *Arch. Int. Pharmacodyn. Thér.* 70: 383–93.

De Duve. C., P. P. De Nayer, M. Oostveldt, and J. P. Bouckaert. 1945. "The Action of Insulin. II. The Hepatectomized and Eviscerated Animal." *Arch. Int. Pharmacodyn. Thér.* 70: 78–98.

De Duve, C., B. C. Pressman, R. Gianetto, R. Wattiaux, and F. Appelmans. 1955. "Tissue Fractionation Studies. 6. Intracellular Distribution Patterns of Enzymes in Rat-Liver Tissue." *Biochem. J.* 60: 604–17.

Hers, H. G. 1963. "Alpha-Glucosidase Deficiency in Generalized Glycogen Storage Disease (Pompe's Disease)." *Biochem. J.* 86: 11–16.

Hers, H. G., J. Berthet, L. Berthet, and C. de Duve. 1951. "The Hexose-Phosphatase System. III. Intracellular Localization of Enzymes by Fractional Centrifugation." *Bull. Soc. Chim. Biol.* 33: 21–41.

Soskin, S., and R. Levine. 1937. "A Relationship between the Blood Sugar Level and the Rate of Sugar Utilization, Affecting the Theories of Diabetes" *Am. J. Physiol.* 120: 761–70.

Sutherland, E. W., and C. de Duve. 1948. "Origin and Distribution of Hyperglycemic/Glycogenolytic Factor of the Pancreas." *J. Biol. Chem.* 175: 663–74.

Tricot, J. P. 2006. "Nobel Prize Winner Chrisian de Duve. From Insulin to Lysosomes." *Hormones* 5: 151–55.

14

Arthur Kornberg

A Giant of Biochemistry

Service in the U.S. Public Health Service and Research with Ochoa

Arthur Kornberg (Figure 17) spent only six months in the Cori department, where he learned the power of enzyme purification. He was the son of Eastern European Jews and grew up in Brooklyn, where he attended Abraham Lincoln High School. He was a precocious student and skipped three years of school. Kornberg entered the City College of New York at age fifteen and graduated at age nineteen with a BS in chemistry and biology. Because of the Depression, jobs were scarce; he decided to enter medical school at the University of Rochester. Of his class of two hundred pre-med students at the City College, he was only one of five who was accepted to medical school. Kornberg supported himself through the first half of medical school on a New York State Regent's Scholarship and the money he had earned during college working as a salesman in a men's store during evenings, weekends, and holidays.

Kornberg enjoyed medical school but considered biochemistry dull because of the descriptive nature of the course. He was denied several research fellowships because of the religious barriers prevalent at that time. Nevertheless, he undertook a project on his own when he realized he had mild jaundice. This involved measuring the blood level of the bile pigment bilirubin in himself, some fellow students, and patients who were normal or exhibited mild jaundice. He continued the project when he was an intern, and his findings were reported in the *Journal of Clinical Investigation*, where it was shown that the defect was due to a reduction in urinary excretion of bilirubin—a benign familial condition now called Gilbert's Syndrome.

Following a one-year internship in Rochester, Kornberg became a commissioned officer in the U.S. Public Health Service and served as a doctor on a naval vessel. His paper on bilirubin came to the attention of the director of NIH, who was concerned about the high incidence of jaundice in servicemen inoculated with yellow fever vaccine. He therefore arranged for Kornberg to be transferred to NIH. Since the focus there was on the role of vitamins in nutrition, Kornberg was assigned to discover why rats that were fed a synthetic diet and a sulfonamide drug developed a blood disorder and died. This didn't happen with the inclusion of a yeast or liver supplement, and the factor was traced to folic acid. Further work indicated that the sulfonamide was blocking the synthesis of folic acid and vitamin K by the intestinal bacteria, and this effect could be offset by administering para-aminobenzoic acid. These findings constituted Kornberg's first paper in the *Journal of Biological Chemistry*.

In 1945, Kornberg was tiring of feeding rats different diets and was becoming excited by the papers on enzymes, coenzymes, and ATP emerging from the laboratories of Otto Warburg, Otto Meyerhof, Herman Kalckar, Fritz Lipmann, and Carl Cori. So he requested a transfer to a laboratory dealing with enzymes and ATP and became an apprentice to Bernard Horecker, who introduced him to cytochrome c, succinic acid oxidase, and oxidative phosphorylation. Horecker was notable fir his elucidation of the pentose phosphate pahway. Kornberg then spent an exhilarating year with Severo Ochoa in New York, where he describes his learning curve as exponential! (Ochoa's biography is presented in a previous chapter.) During this time, he married Sylvy Ruth Levy, a biochemistry student he had met in Rochester. They immersed their nonworking hours in the theater, music, and museums that are the delights of New York City. His project with Ochoa was to purify aconitase, which is the enzyme that converts citric acid to isocitric acid in the citric acid cycle. The enzyme was thought to consist of two components because cis-aconitic acid was an intermediate, but it proved to be a single protein. The enzyme could be assayed in a few minutes by coupling it to isocitrate dehydrogenase and measuring the NADH formed using the Beckman DU spectrophotometer, an instrument that transformed biochemistry.

Kornberg described in his autobiography how he learned from his experience with aconitase the process of enzyme purification and the need to document it meticulously such that it would withstand the scrutiny of an auditor or bank manager! He thought of it in terms of the ascent of an uncharted mountain, with the need to establish successively higher base camps, to withstand fatalities, storms, and other hardships, with gratifying views en route and with the ultimate reward of an exhilarating view from the summit. His second attempt at enzyme purification involved malic enzyme, which converts malic acid to pyruvic acid. Alan Mehler, who acted as Kornberg's tutor—although he was a graduate student and four years younger—assisted him in this.

Research in the Cori Laboratory

Kornberg felt that his experiences in the Ochoa laboratory inculcated a lifelong love of enzymes with an unremitting devotion to their fractionation and purification as a means of resolving biological problems. After spending a year there, he was allowed by NIH to spend six months during 1947 as a commissioned U.S. Public Health Service Officer in the Cori laboratory, which he described as the mecca of enzymology. Kornberg noted that not only were the Coris in the forefront of biochemistry with a focus on enzymes, but they also welcomed refugees from anywhere to join them. Unlike the prevailing culture in American academia, the Coris showed no discrimination toward men or women or Jews or Gentiles. Kornberg commented that their laboratory was the most vibrant place in biochemistry at that time and that scientists flocked there to share in the excitement at the frontiers of intermediary metabolism. The Coris accepted him even though he was a real novice in science and had had no formal research training; they proved to be his most devoted patrons.

Kornberg initially had the audacity to embark on a project to solve the riddle of oxidative phosphorylation, which predictably failed. He then worked with Olov Lindberg, a young Swedish investigator, on the puzzling finding that pyrophosphate was produced during the metabolism of pyruvic acid by liver particles. This was unusual because pyrophosphate was not known to be a cellular constituent. In later work on respiration and coupled phosphorylation in kidney particles, a large stimulatory effect of NAD was observed, and this was traced to AMP formed by its hydrolysis, which was then converted to ATP. This attempt to elucidate the mechanism of oxidative phosphorylation failed because, like most investigators at that time, Lindberg studied the process using soluble enzymes instead of mitochondria.

Kornberg relates an interesting incident that involved Lindberg. There was to be a visit by the eminent Swedish biochemist Hugo Theorell (who is also described in chapter 13 on Christian de Duve), and the Coris decided to arrange a gala dinner on his arrival that would include the social, artistic, and scientific aristocracy of St. Louis. Gerty had a hairdo, put on makeup, and bought a new dress for the first time in anyone's memory. Lindberg went out to the airport to meet the flight, but Theorell was not on it. When Lindberg returned to the department without the guest, Carl asked him to check the telegram, and, sure enough, Theorell was due a week later. Lindberg was devastated, but when Theorell arrived a week later, they all managed to enjoy a laugh over the misunderstanding.

Kornberg describes the Coris as exuding a contagious work ethic, optimism, and a broad view of biological science. He felt that they complemented each other in every way, with Gerty flitting into Carl's tiny office throughout the day with results from the experiments in which she was totally engaged. Carl was calm and analytical in contrast to Gerty, who was agitated and intuitive. When they gossiped about people and events, Carl would exhibit amused concern, whereas Gerty would show intensity and passion. Kornberg recalled an incident when Gerty waved a newly arrived journal exclaiming, "We've been attacked!" However, a careful search of the journal did not reveal the presumed slight. Kornberg notes that Carl's erudition and formidable intelligence could be intimidating and that he could be dismissive.

These characteristics of Carl were also noted by Christian de Duve, Edwin Krebs, and Charles Park in their autobiographies.

Establishment of the Enzyme Section at NIH

Kornberg returned to NIH in 1948, where he set up a biochemical laboratory (Enzyme Section) to be housed in the Laboratory of Physiology in Building 3, which was being renovated. He invited Leon Heppel and Bernard Horecker to join him. Since their planned research would depend heavily on assays using NAD and NADP, they set about to isolate them from sheep liver using a method devised by Otto Warburg, since there were no vendors of such biochemical reagents. This resulted in them having the world's supply of NADP, and when Warburg visited their laboratory, they were able to give the discoverer 25 mg of that coenzyme. The enzyme assays that involved measuring NADH and NADPH used the new Beckman DU spectrophotometer, but there was uncertainty about their true extinction coefficients at 340 nm. When Kornberg and Horecker determined these, their paper making possible quantitative spectrophotometric measurements in reactions involving the nucleotides became one of the most frequently cited papers in the biochemical literature.

The new Enzyme Section provided an exciting and stimulating atmosphere. They held daily luncheon seminars together with Herbert Tabor (from the Laboratory of Pharmacology) to discuss recent papers in the *Journal of Biological Chemistry*. Unbelievably, NIH consisted of only six small buildings at that time, and the focus was on infectious diseases. Kornberg continued working on the enzyme that he and Lindberg had discovered in St. Louis, namely, that which hydrolyzed NAD at the pyrophosphate linkage to produce nicotinamide mononucleotide. He purified the enzyme from potato and found that it cleaved not only NAD, but also all nucleotides with a pyrophosphate bond, and he called it nucleotide pyrophosphatase. He isolated nicotinamide mononucleotide and showed that it could be condensed with ATP to form NAD and pyrophosphate, explaining the origin of this compound in cells. This showed that the reaction was reversible, supporting a vigorous exchange of pyrophosphate with ATP. Kornberg's work led him to the discovery of another enzyme that could synthesize another nucleotide coenzyme—flavin adenine dinucleotide (FAD). Importantly, he recognized that the mechanism of nucleotidyl transfer from a nucleoside triphosphate as exemplified in these reactions was repeatedly involved in the biosynthesis of proteins, lipids, and carbohydrates. His research then focused on the synthesis of the nucleotides involved in DNA synthesis, and he discovered an enzyme (PRPP synthetase) that transferred the pyrophosphate group from ATP to ribose-5-phosphate to produce a novel compound 5'-phosphoribosyl-1-pyrophosphate (PRPP), which later proved to be a key precursor of purine and pyrimidine nucleotides, histidine, tryptophan, and NAD.

Kornberg considered PRPP synthetase one of his favorite enzymes because of the great utility of its products.

Return to Washington University and Discovery of DNA Polymerase

With the encouragement of the Coris, Kornberg moved back to the Washington University School of Medicine in 1953 to become chairman of the Department of Microbiology. Learning from his mentors Ochoa and the Coris that every biochemical process should be amenable to elucidation by enzyme purification, Kornberg undertook the formidable problem of the enzymatic synthesis of DNA. In 1953, James Watson and Francis Crick had discovered the double helix structure of DNA, but it was unclear how it was assembled in the cell. Kornberg started his experiments using ^{14}C-labeled thymidine supplied by Morris Friedkin, a colleague in the Pharmacology Department at Washington University. An extract from *Escherichia coli* was used as an enzyme source because it was more active than mammalian tissues. Although the incorporation of label was very low, treatment with a DNase showed that it was into DNA. Although Kornberg's group was discouraged to hear from Herman Kalckar that Ochoa and Grunberg-Managor had discovered the enzymatic synthesis of RNA, they persisted, but made the error of using ADP as the substrate (which Ochoa used) rather than ATP. When Kornberg repeated his original thymidine incorporation experiments, the results were more encouraging, and he told a postdoctoral fellow, Robert Lehman, about them. Lehman dropped his original project and switched to DNA synthesis. He found that thymidine triphosphate was a much better substrate, and they set out to find the enzyme involved. An advance came when, based on the experiments of the Coris showing that glycogen could act as a primer for its further synthesis by phosphorylase, DNA was added to the reaction mixture. It was later found that DNA acted not only as a primer but also as a template and a source of some missing nucleotides. It took Kornberg and his colleagues (Lehman, Maurice Bessman, and Ernest Simms) about three years to purify the enzyme involved, which they named DNA polymerase, and to show that all four deoxynucleoside triphosphates were required substrates (but see below). Lehman has described the many steps involved in the purification and isolation of the enzyme and the requirements for DNA synthesis. The DNA product reflected the composition of the purine and pyrimidine bases in the template, and sequencing revealed that the two strands of the DNA double helix had opposite polarities—a structural feature that had not been demonstrated previously. The reaction mechanism was shown to involve the nucleophilic attack of the α-phosphorus of the incoming deoxynucleoside triphosphate on the 3' hydroxyl group at the primer terminus to form a new phosphodiester bond. This extended the primer by a single nucleotide with the elimination of pyrophosphate, which was then broken down by a pyrophosphatase thus driving the reaction in the direction of synthesis, This process is repeated sequentially until replication of the DNA template is complete.

Move to Stanford University

In 1957, Kornberg accepted the chairmanship of microbiology at Stanford. The Coris were upset with his decision. For the first time, Carl was angry with him and sputtered, "Where will you go on vacation?" Gerty's reply was: "Carly, maybe we should have gone to California when they asked us." Whatever the scientific climate, the move from the sweltering summers of St. Louis to the cooler air of Northern California was an inducement. Remarkably, two papers describing Kornberg's work on DNA polymerase, which was of fundamental importance in understanding the process of DNA replication, were declined for publication by the *Journal of Biological Chemistry*. Among the critical comments were: "It is very doubtful that the authors are entitled to speak of the enzymatic synthesis of DNA"; "Polymerase is a poor name etc." The reviewers also complained that the deoxynucleoside triphosphates were not adequately characterized. Through the fortunate intervention of John Edsall, newly appointed as editor in chief of the journal, the decision was reversed and the papers were published in revised form in 1958. The next year, Kornberg shared the Nobel Prize with Severo Ochoa for their work on DNA and RNA synthesis, respectively.

Demonstration of the Biological Activity of DNA Polymerase

Although DNA had been synthesized, it had not been shown to be biologically active. This was a great embarrassment to Kornberg, who had spent more than ten years trying to demonstrate activity. A real breakthrough came with the discovery of the ligase enzyme, which catalyzes the joining of linear DNA molecules and the conversion of linear to circular DNA. Kornberg and Mehran Goulian, together with Robert Sinsheimer from CalTech, were able to replicate the single-stranded circle of the phage φX174 using DNA polymerase and seal the complementary strand with ligase. The circular product was then isolated and replicated to produce a circular copy of the original viral strand. Crucially, it was shown to be just as infectious as the original phage DNA. This achievement involved the high-fidelity replication of a 5,000-nucleotide DNA sequence—a formidable achievement that received wide publicity. The Stanford News Bureau called a press conference because of the many inquiries about their paper in the *Proceedings of the National Academy of Sciences*, and it was dubbed the "creation of life in the test tube." A remarkable discovery was that DNA polymerase could proofread and edit its product. This was shown by deliberately preparing some residues at the primer end, which were not matched to the other strand. These mismatched residues were immediately removed—whether or not deoxynucleoside triphosphates were added—to allow chain extension. The process involves the removal of unmatched residues at the 5' end by the 5' → 3' exonuclease domain of the enzyme. This proofreading and editing action reduces errors in the replication process to one in a million, and, with the operation of other mechanisms, the error rate drops to one in a billion—a crucial requirement for the fidelity needed for normal growth and reproduction of organisms.

As is unfortunately common with breakthrough findings, DNA polymerase was called a "red herring" and was charged in a series of editorials in *Nature New Biology* as masquerading as a replication enzyme! This was because an *E. coli* mutant, which appeared to lack the enzyme, multiplied normally. The polymerase was also being recognized as a repair enzyme, and genes were being discovered that indicated that many other proteins were essential for replication. It was even claimed that the discovery was preventing the identification of the "true" enzymes involved in DNA replication. The situation was resolved in an interesting manner.

Kornberg's middle son Tom had been a student of the cello at the Juilliard School in New York, but a hand injury required a career change and a move to Columbia University. There, he was disturbed to hear disparaging remarks about DNA polymerase in his biology course. Despite a lack of laboratory experience, he set out to disprove the accusations against his father and identified within a few weeks another DNA polymerase in *E.coli* that was different from the one his father had discovered. The new enzyme was called DNA polymerase II, and the original polymerase was termed DNA polymerase I. In the course of purifying DNA polymerase II, Kornberg's son found another termed DNA polymerase III. Despite significant differences in structure, it was found that all three enzymes were virtually identical in their mechanisms of DNA synthesis. With the recognition of several forms of the polymerase, the skepticism surrounding the original reports disappeared as did the journal that had promoted the controversy. DNA polymerase III is now recognized to be absolutely essential for DNA replication in *E. coli*, and two more DNA (polymerases IV and V) have been identified in this organism. The situation is more complex in eukaryotes, where at least fifteen polymerases have been identified.

Starting DNA Chains and Replicating DNA

A question arose from the previous work, namely, how does DNA polymerase I start a new chain? Efforts to demonstrate this were met with several years of unproductive results, and the problem turned out to be the damaged DNA being used as a template and primer. When homogeneous single-stranded circular DNA molecules (the chromosomes of the small bacteriophages M13 and φX174) were used, the process of chain initiation could be studied. As described below, it turned out to be rather complex, and, to understand it, one needs some key facts. Unlike DNA polymerase, RNA polymerase can start chains. Surprisingly, DNA polymerase will accept an RNA chain end-matched to a DNA template (although it will not accept ribonucleotides), and the editing function of DNA polymerase (the 5' → 3' exonuclease) can remove such RNA and replace it with DNA. Thus Kornberg had the idea that RNA polymerase could make a short piece of RNA on single-stranded M13, which DNA polymerase could use to start a DNA chain.

To prove this intriguing conjecture, Kornberg used rifampicin, which inhibits RNA polymerase, and showed that the M13 circle was not replicated. However, the system proved to be more complicated since the drug did not interrupt the replication of the *E. coli* chromosome. A torrent of eight new enzymes that affected replication then came from a variety of laboratories, and sorting out their functions became a major undertaking. Over a period of about ten years, Kornberg's group identified and assembled a group of proteins termed the "replisome" that could convert the single-stranded circular φX174 DNA to its double-stranded replicative form. All the components of the replisome were products of *E. coli* genes that were essential for chromosomal replication. In addition to the multi-subunit DNA polymerase III, there was the primosome consisting of a DNA helicase (an enzyme that unzippers the DNA helix) and a primase (an enzyme that lays down the stretch of RNA that attracts DNA polymerase to start a DNA chain). The replisome assembled by Kornberg was able to catalyze the rapid, high-fidelity replication of a single-stranded circular viral genome to yield a double-stranded circular DNA molecule. This led him to undertake a formidable task, namely, to replicate the genome of a bacterium (*E. coli*). This common inhabitant of the intestines had the advantage of reproducing in minutes rather than days and thus had a very active replication system. In addition, Kornberg believed in the universality of biochemical systems between bacteria, fungi, plants, and animals—a belief that would later be borne out when the replication mechanisms in human cells were elucidated.

Before he could commence this undertaking, he had to find out how the replication of a duplex DNA molecule was initiated, propagated, and terminated. A critical element in solving this problem was the use of a template consisting of a double-stranded circular DNA molecule into which the *E. coli* origin of replication (OriC) had been inserted. This enabled the purification of a group of proteins that recognized OriC, altered its structural conformation, and led to its further opening by helicase action. An important member of these initiating proteins was the *E. coli dna*A gene product, which was not part of the replisome. After twelve man-years of effort, Kornberg and his coworkers were finally able to reconstitute the origin-specific replication of the OriC-containing plasmid, and thereby the replication of the *E. coli* chromosome. This monumental achievement is thought to have influenced other scientists to tackle problems of a seemingly intractable nature. It may have influenced another of Kornberg's sons, Roger, to undertake the elucidation of the three-dimensional structure and dynamics of mRNA synthesis by the RNA polymerase complex, an extraordinary accomplishment that won him the Nobel Prize in 2006.

Kornberg's Philosophy and Impact on Recombinant DNA Technology

The discoveries of Kornberg are largely responsible for the massive program of gene sequencing, cloning, and manipulation that is encompassed in recombinant DNA technology. The ensuing biological revolution has transformed biomedical research and led to the production of large quantities of therapeutic proteins such as insulin, erythropoietin, and growth hormones and modified species. In one area alone, DNA polymerase has become of immense importance, because it forms the basis of the polymerase chain reaction (PCR) developed by Kary Mullis and for which he won the Nobel Prize in Chemistry in 1993. This has greatly facilitated cloning of DNA and, through its immense power of amplification, has allowed the determination of DNA from prehistoric species and from microscopic amounts of human tissue. Thus, it has aided in the identification of genetic disorders and in determining the identity of individuals at crime scenes.

Despite his heavy involvement in research, Kornberg found the time to write several books. Classics are his *DNA Synthesis* and *DNA Replication* and his autobiographical *For the Love of Enzymes*. He also drew on his experiences as founder of the DNAX Research Institute to write *The Golden Helix: Inside Biotech Ventures*. His last book, *Germ Stories*, was a collection of poems written for his children and grandchildren. Arising from his love of enzymes were his "Ten Commandments of Enzymology," which were published in *TRENDS in Biochemical Sciences*. These comprise the following: Thou shalt... 1. Rely on enzymology to resolve and reconstitute biological events. 2. Trust the universality of biochemistry and the power of microbiology. 3. Not believe something just because you can explain it. 4. Not waste clean thinking on dirty enzymes. 5. Not waste clean enzymes on dirty substrates. 6. Use genetics and genomics. 7. Be aware that cells are molecularly crowded. 8. Depend on viruses to open windows. 9. Remain mindful of the power of radioactive tracers. 10. Employ enzymes as unique reagents.

Kornberg was a passionate and effective champion of basic, untargeted research, and his strongly held views did not sit well with the protagonists of goal-oriented research and some members of Congress. He defended his position in numerous national addresses and articles, pointing out how arcane his studies of phages and bacteria seemed at the outset. He also described how the discoveries of penicillin, X-rays, polio vaccine, monoclonal antibodies, genetic engineering, and recombinant DNA came from research that had no practical objectives and was not related at the outset to a specific medical problem. He noted that these discoveries did not come from industry or goal-oriented projects, and that the pursuit of curiosity about the basic facts of nature has proven throughout the history of medical science to be the most cost-effective route to successful drugs and devices. He also pointed out that the multibillion-dollar biotechnology industry owed its origin to the enzymes discovered by untargeted research. He was also a great supporter of NIH, which he called his alma mater. He was grateful for the support that NIH had provided at all phases of his career, giving out many millions of dollars for research that had no apparent relevance to practical applications or human disease.

Besides the Nobel Prize, Kornberg won many prestigious awards during his lifetime. These include the National Medal of Science, the Gairdner Award, and the Cosmos Club Award. He was elected to the National Academy of Sciences, the American Academy of Arts and Sciences, the American Philosophical Society, and as a Foreign Member of the Royal Society. In 1965, he served as president of the American Society for Biological Chemistry. He also received honorary doctorate degrees from twelve universities. As a departmental chairman, his prime interest was in research rather than teaching—an attitude that was not appreciated by one of his deans. Experiments were more fulfilling, although he did enjoy a modest amount of formal lecturing. For him, the most rewarding teaching was in the intimate, daily contact with graduate and postdoctoral students. He felt closest to those who shared his devotion to enzymes. He recalled an incident in 1948 when he complained to Sidney Colowick and Oliver Lowry about his failure in purifying an enzyme using a particular method. "I wasted a whole afternoon trying that," whereupon Colowick turned to Lowry and said with mock gravity, "Imagine, Ollie, he wasted a *whole* afternoon."

In comparing the contributions of imagination or hard work to scientific discoveries, Kornberg chose a middle road—hard work with a touch of fantasy. He said he put off writing even though it was an integral part of research, and he felt uneasy when students were writing at their desks rather than working at the laboratory bench. Although in general he found writing scientific papers a chore, he authored some outstanding textbooks. He also said that his duties as a departmental chairman did not unduly intrude on his scientific activities, except for his obligations to the executive committee of the medical school where science and education policy were never discussed. Throughout his career, he made sure that research was his first priority, and he made a point of mixing graduate students with postdoctoral fellows. Furthermore, he arranged that the general biochemistry space was shared by all members of the department, which maximized interaction and collaboration and the sharing of reagents, equipment, and methods.

References

Bessman, M. J., I. R. Lehman, E. S. Simms, and A. Kornberg. 1958. "Enzymatic Synthesis of Deoxyribonucleic Acid. II. General Properties of the Reaction." *J. Biol. Chem.* 233: 171–77.
Goulian, M., A. Kornberg, and R. L. Sinsheimer. 1967. "Enzymatic Synthesis of DNA. XXIV. Synthesis of Infectious Phage Phi-X174 DNA." *Proc. Natl. Acad. Sci. U.S.A.* 58: 2321–28.

Kornberg, A. 1942. "Latent Liver Disease in Persons Recovered from Catarrhal Jaundice and in Otherwise Normal Medical Students as Revealed by the Bilirubin Excretion Test." *J. Clin. Invest.* 21: 299–308.
Kornberg, A. 1989. "Never a Dull Enzyme." *Ann. Rev. Biochem.* 58: 1–31.
Kornberg, A. 1991. "Understanding Life as Chemistry." *Clin. Chem.* 37: 1895–99.
Kornberg, A. 1997. "Science and Medicine at the Millenium." *Biochem. Mol. Med.* 61: 121–26.
Kornberg, A. 1997. "Centenary of the Birth of Modern Biochemistry." *FASEB J.* 11: 1209–14.
Kornberg, A. 2001. "Rembering Our Teachers." *J. Biol. Chem.* 276: 3–11.
Kornberg, A. 2003. "Ten Commandments of Enzymology, Amended." *Trends in Biochemical Sciences* 28: 515–17.
Kornberg, A., F. S. Daft, and W. H. Sebrell. 1943. "Production and Treatment of Granulocytopenia and Anemia in Rats Fed Sulfonamides in Purified Diets" *Science* 98, 20–22.
Kornberg, T., and M. L. Gefter. 1970. "DNA Synthesis in Cell-Free Extracts of a DNA Polymerase-Defective Mutant." *Biochem. Biophys. Res. Commun.* 40: 1348–55.
Kornberg, T., and M. L. Gefter. 1971. "Purification and DNA Synthesis in Cell-Free Extracts: Properties of DNA Polymerase II." *Proc. Natl. Acad. Sci. U.S.A.* 68: 761–64.
Lehman, I. R. 2003. "The Discovery of DNA Polymerase." *J. Biol. Chem.* 278: 34733–38.
Lehman, I. R. 2008. "Historical Perspective: Arthur Kornberg, a Giant of 20th Century Biochemistry." *Trends in Biochemical Sciences* 33: 291–96.
Lehman, I. R. 2010. "Arthur Kornberg 1918–2007." *Biogr. Mem. Natl. Acad. Sci. U.S.A.* 2–20.
Lehman, I. R., M. J. Bessman, E. S. Simms, and A. Kornberg. 1958. "Enzymatic Synthesis of Deoxyribonucleic Acid. 1. Preparation of Substrates and Partial Purification of an Enzyme from Escherichia Coli." *J. Biol. Chem.* 233: 163–70.

15

Hormone Effects on Muscle Carbohydrate Metabolism

Studies with Diaphragm Muscle

Diabetes mellitus causes an increase in blood glucose because of effects on both the liver and muscles. Although in the 1940s information was emerging about the actions of glucagon, insulin, and epinephrine on liver glycogen, the effects of hormones on muscle carbohydrate metabolism were less clear. Carl Cori was no longer active in the laboratory in 1947, but he supervised an extensive series of studies of the effects of insulin, epinephrine, pituitary, and adrenal cortical hormones on the uptake and metabolism of glucose by muscle, utilizing intact animals and isolated muscle preparations as experimental systems. As described in chapter 2, devoted to Sidney Colowick, these studies were preceded by some experiments using muscle extracts, which indicated that the hexokinase activity in extracts from various tissues could be directly inhibited by an anterior pituitary extract and that insulin, added directly to the extracts, could relieve the inhibition. As noted previously, to the great dismay of Carl and Sidney Colowick, who were co-investigators in the study, these results could not be reproduced. This study, which was published, greatly distressed Carl, who, as director of the laboratory, aspired to the highest standards of scientific integrity. However, he never withdrew the hexokinase theory of insulin action. He finally accepted the finding that the hormone increased the transport of glucose across the cell membrane, but felt that it was secondary to some change in the intracellular metabolism of glucose.

Carl's studies of hormone effects on glucose metabolism in muscle were reported starting in 1947, and his group employed diaphragm muscles prepared from rats treated in various ways. Some of these rats were obtained from Anheuser-Busch, Inc., which illustrates that, at this time, the company produced rats as well as beer. Many associates were involved in the experiments, including Mike Krahl, Charles Park, Joe Bornstein, David Kipnis, Marvin Cornblath, Ernst Helmreich, David Brown, William Daughaday, and William Danforth. These were almost all young physicians who had recently joined the laboratory and who later had impressive careers in academic medicine. The careers of Park, Kipnis, Helmreich, Brown, Daughaday, and Danforth are described in detail in later chapters with appropriate references. The focus of the research was on the effects of diabetes, insulin, and hormones from the pituitary and adrenal glands on the entry and phosphorylation of glucose in diaphragm muscle both in vivo and in vitro. Their research utilized a variety of approaches, such as the study of glucose itself, which is both transported and phosphorylated; pentoses, which are 5-carbon sugars that are transported but not metabolized; and 2-deoxglucose, which is transported and phosphorylated but not further metabolized. Through the use of these different sugars, the effects of the hormones on transport could be distinguished from effects on phosphorylation. The research involved animals that had been treated in various ways, for example, rats that were made diabetic with alloxan, rats that had had their adrenal glands removed (adrenalectomized), rats that had had their pituitary glands removed (hypophysectomized), and rats with combinations of these procedures. The research also included frog muscles, which were forced to contract by electrical stimulation in situ. In addition to these procedures, the muscles were incubated with insulin, epinephrine, and extracts from the adrenal glands (adrenal cortical steroids) and the anterior pituitary glands (growth hormone) to test for their effects.

The experiments are described in more detail in the chapters devoted to Charles Park, Ernst Helmreich, David Kipnis, David Brown, and William Danforth. In short, the experiments represented an exhaustive study of the hormonal regulation of muscle carbohydrate metabolism, and the results indicated that insulin acted at the membrane to stimulate the entry of glucose into the muscle cell. Contraction of muscles induced by electrical stimulation also increased glucose penetration. On the other hand, growth hormone and adrenal cortical hormones, acting in a synergistic manner, antagonized the action of insulin on glucose uptake, and it was shown that their effects were not exerted on the transport of glucose, but on its phosphorylation. Diabetes was found to reduce the phosphorylation of glucose, and removal of the adrenal glands and/or the pituitary gland relieved the inhibition. In a separate study, epinephrine was observed to inhibit glucose uptake, but this was not due to reduced glucose transport but to the fact that the stimulation of glycogenolysis caused an increase in glucose-6-phosphate, which inhibited glucose phosphorylation by hexokinase.

All of these findings were highly relevant to the disease of diabetes and explained how insulin lowered the blood glucose through its action on muscle. The results also explained how alterations in the secretion of adrenal cortical hormones and growth hormone could affect the severity of the disease. In addition, they provided a mechanism for how exercise could improve glucose utilization in diabetes, namely though its stimulatory effects on muscle glucose uptake, and how epinephrine released during stress counteracts the effect of insulin on this process. Although Carl did not include his name on some of the publications emerging from this work (because he didn't feel his input was sufficient), his guiding influence and rigorous standards for designing the experiments, ensuring the validity of the results, and carefully interpreting them are clearly evident. As noted above, a large number of gifted young researchers participated in the experiments, and individual chapters are devoted to the most prominent ones.

The other investigators to whom chapters are not devoted nevertheless had significant careers, which are now briefly described. Joe Bornstein was a colorful Australian who combined great enthusiasm with a tendency to exaggerate. He worked with Charles Park to show that serum from diabetic rats inhibited glucose uptake by the rat diaphragm and conducted "midnight experiments" in the Cori laboratory which showed that insulin stimulated glucose transport into diaphragm muscle rather than phosphorylation. These findings were contrary to the Coris' beliefs and were not published at that time. When Park later moved to Vanderbilt Medical School, Bornstein joined him to demonstrate that insulin increased the concentration of free glucose in the diaphragm. Such a result would be expected if the hormone stimulated the transfer of glucose across the cell membrane, but not if it increased its phosphorylation. Bornstein returned to the University of Melbourne in Australia and later became the founding chairman of biochemistry at Monash University. His studies of the effects of polypeptides derived from growth hormone were controversial.

After his work with Carl Cori, Mike Krahl went to the Physiology Department at the University of Chicago and remained there for the rest of his career. His focus became the stimulatory effect of insulin on protein synthesis in the diaphragm, which, surprisingly, was found to be independent of the stimulation of glucose uptake. In another extensive series of studies, Krahl explored the properties and hormonal control of hexokinase in rats.

Marvin Cornblath was another member of the Cori group. Cornblath later focused his career on the regulation of the blood glucose level in the newborn infant, which is characteristically low. After periods spent at hospitals in Chicago, he was appointed to the chair of pediatrics at the University of Maryland, where he combined clinical investigations with basic research. The clinical research involved studies of metabolite and hormonal levels in underweight and normal newborn infants and those from diabetic mothers. The basic research involved studies of enzymes involved in gluconeogenesis and ketone body metabolism in the livers and brains of newborn and adult animals.

References

References are only provided for research not presented in other chapters.

Bornstein, J., and C. R. Park. 1953. "Inhibition of Glucose Uptake by the Serum of Diabetic Rats." *J. Biol. Chem.* 205: 503–11.

Cornblath, M., S. H. Wybregt, G. S. Baens, and R. I. Klein. 1964. "Symptomatic Neonatal Hypoglycemia. Studies of Carbohydrate Metabolism in the Newborn Infant." *Pediatrics* 33: 388–402.

Krahl, M. E. 1964. "Stimulation of Peptide Synthesis in Adipose Tissue by Insulin without Glucose." *Am. J. Physiol.* 206: 618–20.

Park, C. R., J. Bornstein, and R. L. Post. 1955. "Effect of Insulin on Free Glucose Content of Rat Diaphragm in vitro." *Am. J. Physiol.* 182: 12–16.

Pilkis, S. J., R. J. Hansen, and M. E. Krahl. 1968. "Hormonal Control of Hexokinase in Animal Tissues." *Diabetes* 17: 300–01.

Four Nobel Laureates honoring Carl Cori on his 60th birthday. From left, Arthur Kornberg, Severo Ochoa, Carl Cori and Luis Leloir. Photo courtesy of Becker Medical Library of Washington University School of Medicine.

Carl Cori in his office. Photo courtesy of Becker Medical Library.

Gerty Cori unusually in a dress with necklace. Photo courtesy of Becker Medical Library.

Carl and Gerty with the Nobel Laureates Joseph Erlanger (between them) and Arthur H. Compton (Chancellor of Washington University) on right. Photo courtesy of Becker Medical Library.

Carl with Gerty at her desk in their laboratory. Photo courtesy of Becker Medical Library.

Carl with Gerty filtering an extract in their laboratory. Photo courtesy of Becker Medical Library.

Sidney Colowick at his desk at Vanderbilt University. Photo courtesy of Eskind Biomedical Library of Vanderbilt University.

Herman Kalckar. Photo courtesy of the National Library of Medicine.

Severo Ochoa. Photo courtesy of the National Library of Medicine.

Carl Cori with Severo Ochoa with his characteristic cigarette. Photo courtesy of Becker Medical library.

Luis Leloir. Photo courtesy of the National Library of Medicine.

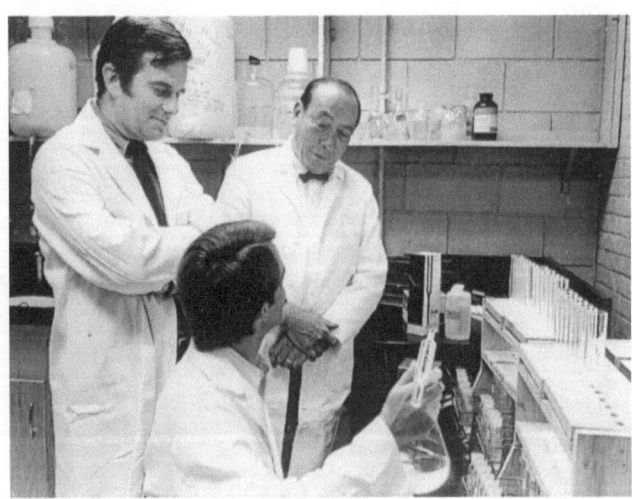

Earl Sutherland (with hands crossed) in his laboratory at Vanderbilt University with Joel Hardman (standing) and Ted Chrisman seated. Photo courtesy of Joel Hardman.

Victor Najjar with unnamed colleague. Photo courtesy of Jennifer Najjar.

Edwin Krebs. Photo courtesy of Jackie Corbin.

Mildred Cohn. Photo courtesy of Becker Medical Library.

Three Nobel Laureates with a painting of Alfred Nobel. From left, George Palade, Albert Claude and Christian de Duve. Photo courtesy of Christian de Duve.

Arthur Kornberg. Photo courtesy of the National Library of Medicine

Charles Park. Photo courtesy of Jackie Corbin.

Jane Park (seated) with Charles Park. Photo courtesy of Eskind Biomedical Library.

William Daughaday in his laboratory. Photo courtesy of Becker Medical Library.

Luis Glaser at his microscope. Photo courtesy of Becker Medical Library.

Ernst Helmreich at the Würzberger Biozentrum. Photo courtesy of Ernst Helmreich with kind permission of Springer Science+Business Media.

Carl Frieden in his laboratory. Photo courtesy of Becker Medical Library.

David Kipnis in his laboratory. Photo courtesy of Becker Medical Library.

William Danforth with students in the physiology laboratory. Photo courtesy of Becker Medical Library.

William Danforth with Carl Cori on his 80th birthday. Photo courtesy of Becker Medical Library.

16

Charles Park

Aristocratic Physiologist

Family Background and Early History

Charles Rawlinson (Rollo) Park has a distinguished heritage, being descended from the eighteenth-century theologian Jonathan Edwards on his father's side and from Sir Henry Rawlinson on his mother's side (Figure 18). Edwards was known for his defense of Calvinist theology and later became president of the College of New Jersey (now Princeton University). His reputation was somewhat tarnished by the fact that he was the grandfather of Aaron Burr, who shot Alexander Hamilton in a duel and was later tried for treason. Rawlinson was a distinguished diplomat, orientalist, and archeologist who gained fame for deciphering the Behistun cuneiform inscriptions that dated from the time of King Darius of Persia. This is comparable to Chompollion's deciphering of the Egyptian hieroglyphics. Rawlinson's son was a general in the British Army. However, his son's career is not so laudatory since he was involved in planning the Battle of the Somme, one of the great disasters of World War I.

Park was born in Baltimore on March 2, 1916, and barely survived anaphylactic shock when he was fed cow's milk. His father, Edwards A. Park, was a noted pediatrician and a pioneer in the field who became chairman of the Pediatrics Department at Johns Hopkins, a prized position he held from 1927 to 1969. When Park was a teenager, he was sent to Göttingen to stay with a German family for a year because his parents thought he was not mature enough for college. He was accepted into Harvard, but his performance was mediocre because his heart and mind were not in it, and he barely graduated in 1937.

Despite his poor academic record at Harvard, Park was accepted into Johns Hopkins Medical School, and this was helped by the fact that the interviewer knew his father. Park did a research project with Barry Wood, an assistant in the Department of Medicine who had had a spectacular athletic career at Harvard, earning nine varsity letters as a quarterback and captain of the football team, a hockey center, a star shortstop, and a first baseman. He was a unanimous choice as all-American quarterback and earned a tenth letter in tennis. All of these athletic endeavors did not prevent Wood from graduating summa cum laude. The work that Park did with him showed that *para*-aminobenzoic acid (PABA) was an essential factor for bacterial growth. It was an auspicious entry into the world of investigative medicine because sulfonamide drugs exert their antibacterial activity by acting as antagonists of PABA.

Park graduated second in his medical class, and, after an internship at Hopkins, he went to Boston, where he was appointed assistant resident and then chief resident at the Peter Bent Brigham Hospital. Following his time at the Brigham, Park served in the U.S. Army Medical Corps. He was sent to Fort Knox, Kentucky, not to guard the gold but to study thermoregulation during the acclimatization of military personnel exposed to a desert-type environment. For this, the enlisted men had to endure the measurement of rectal temperatures and walking on a treadmill in the nude.

Research in the Cori Laboratory

At this point in his life, Park was undecided between clinical work and research, but his father and others believed that biochemistry would become a major research area in the future and encouraged him to apply to the Cori laboratory in 1947. He was never interviewed but thinks that Barry Wood, who was now the chairman of medicine at age thirty-two (six years after obtaining his MD) encouraged the Coris were encouraged to accept Park. As noted above, Wood was gifted intellectually as well as athletically, and Park was very impressed with him. Park described Wood as remarkably nice and his wife as the most attractive woman he had ever met!

Park was very impressed with Carl Cori at their first meeting and felt he was friendly, but austere. On the other hand, he had no interaction with Gerty, who basically ignored him. Park did not at first participate in the Coris' formidable luncheon seminar series, which were *de rigueur* for most of the students and fellows. Park had monthly meetings with Carl, who assigned him a project to explore the action of insulin. At that time, there was great controversy as to whether the major action of insulin was on the liver or muscle. Park looked at the effects of insulin and other hormones on the uptake of glucose by muscle using the rat diaphragm preparation. As noted in the previous chapter, other fellows worked with him on the project, and most of them achieved distinction in academic medicine in the United States and elsewhere.

Bernardo Houssay, the Argentinean physiologist, had won the Nobel Prize with the Coris because of his in vivo investigations of the role of the anterior pituitary gland in diabetes. So Carl was interested in following the role of the pituitary in the regulation of glucose uptake in muscle. Park, working with another fellow, Mike Krahl, studied this using the rat diaphragm preparation and found that glucose uptake was increased by removal of the pituitary gland (hypophysectomy) and reduced by injection of a pituitary extract. Carl interpreted these findings as being due to effects on glucose phosphorylation, as noted in the previous chapter. However, Park, in collaboration with Joe Bornstein, carried out experiments in which the effect of insulin on the intracellular concentration of glucose was measured in the diaphragm. These experiments, which were conducted beyond the purview of the Coris, showed clearly that the effect of insulin was exerted on the transfer of glucose across the cell membrane and not on glucose phosphorylation. Park and Bornstein were excited by this finding, and they showed the data to the Coris, who took the results home and studied them over the next few days. However, Park and Bornstein were dismayed by Carl's response, who, with his guttural accent, stated: "Gerty and I have examined your results and we do not believe them!" The reason for their negative response was because they were still wedded to the idea that insulin acted on hexokinase. These important findings were never published.

Move to Vanderbilt Medical School

In 1952, Park heard that the chairmanship of physiology was open at Vanderbilt Medical School. The search committee invited him after the position had been offered to four candidates, all of whom had turned it down. He was undecided about a career in clinical medicine or in basic medical science. In any event, he went to Nashville to see about the job. He was tempted by the job, because it would mean an increase in his salary from $5,000 as a postdoctoral fellow to $10,000 as a chairman. Park took the job, and when he arrived, he found the department in a terrible state. It was dirty and there was no equipment for his type of research. The total space was 5,000 square feet, but half was given over to medical student teaching. The faculty comprised Robert Post, whose later claim to fame would be his work on Na,K-ATPase, an enzyme system that transports sodium and potassium ions across the cell membrane and is important in the regulation of the ionic composition and membrane potential of cells. Another member was H. C.(Ray) Meng, who worked on the clearing factor, an enzyme that hydrolyses lipoproteins in the blood. Meng later achieved fame through the development of parenteral nutrition, which is a means of providing complete nutrition intravenously to patients who are unable to take this by mouth. From these humble beginnings, Park built one of the best physiology departments in the United States. One of his first recruits was Jane (Janey) Harting, who had also been in the Cori department. Park recognized that the most effective way to recruit her was to marry her. It worked! Park, who had never been trained in physiology, apart from medical school, was faced with teaching the medical class, and he found this was a tough assignment in which he barely kept ahead of the students.

Studies of Glucose Transport

Despite the limitations of the available equipment, Park began a research program based on the effect of insulin to stimulate the transport of glucose into muscle cells. The idea that insulin affected the transfer of glucose across the cell membrane was suggested as early as 1939 by Ejnar Lundsgaard and reinforced by in vivo studies by Rachmiel Levine working at a research institute in Chicago. Lundsgaard had been Herman Kalckar's PhD mentor in Copenhagen. Park and his associates extended this work by measuring the free glucose concentration in several tissues obtained from rats treated with insulin. They found marked increases in the glucose content of diaphragm and heart muscle, and that this was inhibited in muscle from diabetic rats. They realized that the results could not be attributed to a stimulatory effect of insulin on hexokinase since this would have led to a decrease in the concentration of free glucose. In a follow-up study, they added the hormone directly to rat diaphragms, and, again, this increased the concentration of free glucose in the muscle. They also measured the phosphorylation of glucose, and no effect of insulin was observed.

Their work was extended in an impressive series of studies which were written up in five classic papers in the *Journal of Biological Chemistry* in 1961. The first utilized the isolated, perfused rat heart as the experimental system to study the effects of insulin and anoxia on the kinetics of the extracellular transfer, membrane transport, and intracellular phosphorylation of glucose. They found that glucose transport through the muscle cell membrane was the major rate-limiting step for glucose uptake and that insulin added to the perfusion buffer accelerated this without any effect on glucose phosphorylation. They also induced anoxia by perfusing the hearts with N_2-CO_2 instead of O_2-CO_2, and this was found to increase both glucose transport and phosphorylation.

The next series of experiments employed hearts from alloxan-diabetic rats. As expected, glucose uptake was depressed due to reduced membrane transport. Addition of insulin in vitro increased the transport, but the response was slower than normal and relatively insensitive to physiological concentrations of the hormone. Importantly, they found that glucose phosphorylation was depressed in the diabetic hearts. Then came a series of experiments that studied the effects of growth hormone, cortisol, and insulin as well as hypophysectomy and adrenalectomy. The major effect of growth hormone was to reduce the sensitivity of glucose transport to stimulation by insulin. Neither hypophysectomy nor adrenalectomy reversed this, but they did relieve the reduced glucose phosphorylation, and this was restored by either growth hormone or cortisol treatment in vivo.

The next experiments explored the basis for the effects of anoxia and diabetes on the hexokinase activity The results indicated that this enzyme was inhibited in the diabetic hearts because the level of glucose-6-phosphate (which exerts an inhibitory effect on the enzyme) was increased. Measurements of intermediary metabolites involved in the glycolytic pathway indicated that the increase in glucose-6-phosphate was attributed to depressed activity of phosphofructokinase, an enzyme involved in the further metabolism of glucose-6-phosphate. Additional work showed that the reduced activity of this enzyme observed in diabetic hearts could be relieved by hypophysectomy and restored by growth hormone and cortisol. In contrast, anoxia caused a decrease in glucose-6-phosphate, which could be attributed to an increase in the activity of this enzyme. The changes in the diabetic hearts were later attributed by Philip Randle's group at Cambridge University to changes in fatty acids and their metabolites in these hearts. In the case of anoxia, the increase in the activity of phosphofructokinase was attributable to changes in the levels of adenine nucleotides. Because the function and metabolism of the heart are closely related to the supply of oxygen and fuel, principally fatty acids, these studies were of fundamental importance in understanding the function of this organ in health and disease.

Efforts were then directed to characterize the system responsible for the transport of glucose across the muscle cell membrane. Earlier work had shown that the mechanism was freely reversible and accelerated in both directions by insulin and anoxia. The experiments involved using the phenomenon of transport counterflow to identify the existence of a mobile carrier as distinct from diffusion or a fixed carrier.

Howard Morgan and David Regen were important participants in the studies utilizing the perfused heart preparation. Morgan was an obstetrician who had graduated from Johns Hopkins at age twenty-one. He was based at an army facility in Clarksville, north of Nashville, and was getting tired of delivering babies from healthy military wives. He went to Vanderbilt looking for research possibilities and ended up working with Park for almost ten years. Morgan was a gifted experimentalist who subsequently became the chair of physiology at the Pennsylvania State Medical School at Hershey. He became president of the American Physiological Society and also of the American Heart Association. David Regen started medical school at Vanderbilt and then switched to the PhD program. He then went to Munich as a postdoctoral fellow to study cholesterol synthesis in the laboratory of Feodor Lynen. Regen returned to Park's department in 1963 and resumed working with Morgan. He claimed his greatest interest was music, which ranged from baroque, for which he constructed an harpsichord, blue grass, for which he played a dobro in a band he had organized, and gospel music, which he performed in local churches. Tetsuro Kono was another important member of the department who studied glucose transport and insulin action. He was born in Tokyo and had vivid memories of the bombing of the city by Allied planes during World War II. Kono obtained his PhD from Tokyo University and spent a postdoctoral fellowship with Sidney Colowick. He then moved to the Park department, where his greatest contribution was the demonstration of the mechanism by which insulin stimulated glucose transport. This was that insulin induced the translocation of glucose carriers from an intracellular site to the plasma membrane. Sam Cushman reached a similar conclusion by working independently at NIH.

Arrival of Earl Sutherland

The department received a boost in 1963 with the arrival of Earl Sutherland and his group from Cleveland. Cyclic AMP became a focus of research for many departmental members. Sutherland entered into many productive interactions with Park's group and also that of Grant Liddle, who was chairman of the Department of Medicine. The discoveries made during the earlier phase of Sutherland's career are documented in the chapter devoted to him, and the following will summarize the work at Vanderbilt. The first significant finding was the demonstration that cyclic AMP was increased in the livers of diabetic animals. The lead researcher in these studies was Leonard (Jim) Jefferson, a graduate student who impressed everyone by driving a Cadillac. Work conducted by Bill Butcher, a member of Sutherland's group, showed that insulin added in vitro could lower the level of cyclic AMP in adipose tissue and liver when it was raised by epinephrine or glucagon. The next phase of the work was directed to defining the enzyme(s) that was the target of insulin action (i.e., adenylyl cyclase or cyclic AMP phosphodiesterase). No consistent effects on the cyclase could be observed, however activation of the phosphodiesterase could be shown. The latter work was mainly due to a colorful New Zealander, J. G. T. Sneyd, who was nicknamed "Sam" after Sam Snead, the golfer. As noted in chapter 7, Sutherland carried out extensive and unique studies of cyclic AMP metabolism in human subjects with emphasis on the effects of hormones.

With Sutherland's worsening illness, much of the day-to-day running of his laboratory devolved to Joel Hardman, who had obtained his PhD from Emory University. Hardman led the work on cyclic GMP and guanylyl cyclase. He recognized that this enzyme existed in soluble and particulate forms. One of his major duties was to supervise the work of two outstanding young investigators, David Garbers and Joe Beavo. Beavo made a detailed study of cyclic AMP phosphodiesterase and deduced, correctly, that there were multiple isoenzymes. He took a position at the University of Washington, where he identified and cloned most of these. Garbers had a spectacular career focused mainly on the role of membrane-associated guanylyl cyclases that mediated a variety of biological events. After time at Vanderbilt, Garbers joined Alfred Gilman's Department of Pharmacology at the University of Texas Southwestern Medical School. His first studies involved the molecular basis of fertilization, where he identified peptides released from eggs that interacted with receptors on sperm that were coupled to guanylyl cyclase, leading to increased motility of the sperm. He expanded this work to identify different receptors for atrial and brain natriuretic peptides throughout the body and a receptor for diarrhea-causing enterotoxins in the intestine. In another important area, Garbers showed that receptors in the olfactory sensory epithelium and the eye were linked to guanylyl cyclase.

Studies of Liver Carbohydrate Metabolism

Another person who joined the Park laboratory a few months after the arrival of Earl Sutherland was the author of this monograph, John Exton, who, like Sneyd, had graduated from the University of Otago in Dunedin, New Zealand. At the time of his arrival, Park was utilizing the isolated perfused rat liver preparation to study insulin action on the liver, but the effects were erratic due to problems with the apparatus. Together with Howard Morgan, Exton redesigned the system and initiated a series of studies on hormone effects on the liver. The first experiments were on the hormonal regulation of gluconeogenesis, which is the process by which lactic acid, glycerol, and many amino acids are converted to glucose in the liver. Gluconeogenesis plays an important role in the early stages of starvation by supplying glucose to the brain at a time when it cannot utilize other substrates. It also contributes to the increased level of blood glucose seen in diabetes.

The first study examined the rates of glucose production by the liver perfused with physiological concentrations of various substrates (lactate, pyruvate, glycerol, alanine, and other gluconeogenic amino acids). The results showed that gluconeogenesis could be regulated simply by the supply of these substrates to the liver. They also indicated the site of the rate-limiting step in the gluconeogenic pathway from lactate. The next series of papers looked at the effects of glucagon, epinephrine, norepinephrine, and cyclic AMP on gluconeogenesis and showed that this nucleotide was a major regulator of the process. The following study utilized measurements of intermediary metabolites in the gluconeogenic pathway to identify more precisely the rate-limiting reaction for lactate gluconeogenesis and the site at which cyclic AMP acted. This entailed a massive number of analyses and showed that the rate-controlling reaction was located at the substrate cycle between pyruvate and phospho*enol*pyruvate. This was later confirmed by the Swedish investigator Lorentz Engström on the basis of enzyme measurements. Another physiologically significant finding was that the effects of glucagon and epinephrine on glucose production by the liver could be antagonized by insulin. Importantly, the antagonism was seen with concentrations of the hormones found in portal blood.

Another member of the Exton group was Larry Mallette, a gifted MD, PhD student who focused on amino acid metabolism in the perfused liver. He found that the liver produced glucose and urea when perfused with amino acids and that glucagon and epinephrine were stimulatory. Furthermore, studies with nonmetabolizable amino acids showed that glucagon stimulated their inward transport. With the arrival of Michio Ui from Japan, studies of the effects of glucagon on amino acid metabolism were extended to include isotopic tracer studies and confirmed that the pyruvate-phospho*enol*pyruvate substrate cycle was also a major site of glucagon action. Ui was a compulsive worker who never took a holiday and literally ran around the laboratory to save time. When he returned to Japan, he worked on a toxin from *Bordetella pertussis* that stimulated insulin secretion. This was traced to the ability of the toxin to inactivate G_i an inhibitory G protein. The toxin then became a useful tool to explore G protein functions in cells. Because of this and other findings, Ui became a leader in biochemical science in Japan.

With Exton moving on to other independent research projects, studies of gluconeogenesis per se in the Park laboratory now became mainly the purview of Simon Pilkis. Pilkis had earned an MD, PhD at the University of Chicago under Mike Krahl, who had worked in the Cori laboratory. Pilkis confirmed the observation of the Swedish investigator Lorentz Engström that glucagon inhibited pyruvate kinase because of the phosphorylation of the enzyme by PKA. This supported the view that the resulting inhibition of glycolytic flux was a mechanism by which the hormone stimulated gluconeogenesis. The field of research on glycolysis and gluconeogenesis was shaken up by the discovery of a new dual-function regulatory molecule, fructose-2,6-bisphosphate. There was some dispute as to who discovered it first, with the strongest claimant being the group of Henri-Géry Hers in Brussels. Nevertheless, it became the focus of Pilkis' research. The molecule was found to stimulate glycolysis by activating the glycolytic enzyme 6-phosphofructo-1-kinase and to inhibit gluconeogenesis by reducing the activity of the gluconeogenic enzyme fructose-1,6-bisphosphatase. Pilkis initiated a vigorous program to understand the formation and action of fructose-2,6-bisphosphate, which was termed a master switching signal. Most surprising was the finding that its synthesis and breakdown were found to be both catalyzed by a single unique bifunctional enzyme. Another surprise was that glucagon caused a dramatic drop in the concentration of fructose-2,6-bisphosphate in the liver, thus providing another mechanism for stimulating gluconeogenesis and inhibiting glycolysis. The decrease in fructose-2,6-bisphosphate was later attributed to the action of PKA to phosphorylate a single serine residue in the bifunctional enzyme. Intriguingly, this single change resulted in an increase in fructose-2,6-bisphosphatase activity and a decrease in 6-phosphofructose-2-kinase activity. From an enzymological point of view, these findings were extremely novel. In 1986, Pilkis moved to the State University of New York at Stony Brook as chairman of physiology and biophysics, where he continued working on the bifunctional enzyme. He moved to the University of Minnesota and sadly died soon after.

The findings of Exton and Pilkis were made using isolated livers and enzymes. With such preparations, it is important to confirm the findings using in vivo systems. This was accomplished by Alan Cherrington, another researcher with a strong interest in the regulation of hepatic metabolism, who had obtained his PhD at the University of Toronto. His forte was carrying out in vivo metabolic studies in intact animals, and these led him to prominence in the field of diabetes. The principal experimental system was the dog in which catheters had been implanted in a variety of vessels for sampling or infusion of hormones or substrates. His initial studies explored the interplay between glucagon and insulin in the control of hepatic glycogenolysis and gluconeogenesis and provided in vivo confirmation of Exton's findings using the perfused liver. Cherrington's experiments were greatly enhanced by the use of somatostatin, which inhibits the secretion of endogenous glucagon and insulin, thus allowing the infusion of these hormones to produce controlled levels in the blood. The studies also allowed the separation of hormone effects on gluconeogenesis in the liver from those due to the release of substrates from peripheral tissues. Another advantage was that the approach permitted the blood glucose level to be maintained constant by glucose infusion in studies involving insulin so that secondary effects due to hypoglycemia could be avoided.

Other work explored the effects of epinephrine and norepinephrine on hepatic glucose output. The focus was on the roles played by direct effects on the liver versus changes in the supply of substrates from peripheral tissues. Another issue was the definition of the adrenergic receptors involved. In addition to exploring the role of sympathetic nervous supply to the liver, there were extensive studies of the role of the vagus nerve. All of this work defined the roles of sympathetic and parasympathetic innervation to the liver in the regulation of glycogenolysis and gluconeogenesis. Although the focus in Cherrington's work was on carbohydrate metabolism, he also explored the regulation of ketone body production and alanine metabolism in the liver. Some studies were extended to man, and the effects of fasting, cortisol, and exercise were studied in the dogs. All of this work was highly relevant to diabetes, and Cherrington achieved prominence in the American Diabetes Association and became its president in 2004–2005.

Studies of Protein Kinases and Phosphodiesterases

Another person of note in the Park department was Jackie Corbin, who called himself a hillbilly since he was raised in the Appalachian Mountains between North Carolina and Tennessee. He gained a PhD with Park and then then did a postdoctoral with Edwin Krebs at a time (1968) when the significance of PKA was just emerging. He returned to Vanderbilt to begin a strong program based on the hormonal regulation of PKA and the mechanism by which it is activated by cyclic AMP. This was extended to include cyclic GMP-dependent protein kinase (PKG) and the phosphodiesterases responsible for the breakdown of cyclic AMP and cyclic GMP. Corbin worked in part on this with Sharron Francis, a long-term collaborator. Although the crystal structures of the kinases were not known, it was realized that they were made up of regulatory and catalytic subunits. Corbin showed that cyclic AMP bound to two separate sites on the regulatory subunit of PKA and that activation of the enzyme involved dissociation of the regulatory and catalytic subunits. In contrast, although PKG had a similar structure, activation did not involve this dissociation.

Corbin and Francis engaged in studies of the various isoforms of phosphodiesterase, and one of them, termed PDE5, was found to bind cyclic GMP at a site different from the catalytic site, which selectively hydrolyzed cyclic GMP. Its tissue distribution was found to be relatively limited, that is, in vascular smooth muscle in certain places. Since there was evidence that cyclic GMP was involved in the relaxation of smooth muscle, it was suggested that inhibitors of PDE5 could be useful in treating hypertension. One of these inhibitors, sildenafil (Viagra), was found to be very potent, and it and a related drug, tadalifil (Cialis), were tested as antihypertensive agents in clinical trials. In one of these trials with sildenafil, it was found that men were experiencing improved sexual function because the drug was alleviating erectile dysfunction. This was because the drug's effect to inhibit cyclic GMP breakdown and thus cause relaxation of vascular smooth muscle in the corpus cavernosa resulted in swelling of the penis. A similar effect on the blood vessels of the lung resulted in alleviation of pulmonary hypertension. Thus, from Corbin's basic biochemical studies of PDE5 emerged a blockbuster type of drug that not only alleviates a major problem in elderly men, but is also useful in the treatment of pulmonary hypertension.

Studies on Fatty Acid Transport

Because of their lipophilic nature, it was assumed for many years that fatty acids would cross the cell membrane without the involvement of a carrier system. Working with Nada Abumrad, a postdoctoral fellow from Lebanon, the Parks utilized a rigorous kinetic analysis of fatty acid uptake in isolated fat cells to prove the existence of a membrane transport mechanism. Critical elements of their approach were the measurements of initial rates of uptake; the absence of glucose in the medium so that the fatty acid inside the cells remained unesterified; and the use of phloretin, which inhibited both influx and efflux of fatty acid. The use of phloretin enabled abrupt termination of these fluxes and also washing of the cells without loss of unesterified fatty acid. Later work showed a specificity for fatty acids of different chain lengths and the need for a free carboxyl function for recognition by the transporter. Other work showed inhibitory effects of stilbene compounds and also the labeling of a protein with a radioactive stilbene compound, indicating the protein nature of the transporter. Hormonal regulation of the transport system was revealed when it was shown that catecholamines, especially β-adrenergic agonists and cAMP analogs, could activate fatty acid transport in adipocytes.

Rollo Park was recognized for his contributions by election to the National Academy of Sciences and receipt of the Banting Award from the American Diabetes Association. He was also a counselor of the National Heart Institute and served on the medical advisory boards of the Juvenile Diabetes Foundation, Life Insurance Medical Research Fund, and the Howard Hughes Medical Institute. He also served as vice president of the American Society of Clinical Investigation and president of the Association of Chairmen of Departments of Physiology. Park retired from the chair at Vanderbilt in 1984.

References

Abumrad, N., R. C. Perkins, J. H. Park, and C. R. Park. 1981. "Mechanism of Long Chain Fatty Acid Permeation in Isolated Adipocytes." *J. Biol. Chem.* 256: 9183–91.

Cherrington, A. D., J. L. Chiasson, J. E. Liljenquist, A. S. Jennings, U. Keller, and W. W. Lacy. 1976. "The Role of Insulin and Glucagon in the Regulation of Basal Glucose Production in the Postabsorptive Dog." *J. Clin. Invest.* 58: 1407–18.

Chu, C. A., D. Sindelar, D. W. Neal, E. J. Allen, P. Donahue, and A. D. Cherrington. 1997. "Comparison of Direct and Indirect Effects of Epinephrine on Hepatic Glucose Production." *J. Clin. Invest.* 99: 1044–56.

Corbin, J., and S. H. Francis. 2003. "Molecular Biology and Pharmacology of PDE-5-Inhibitor Therapy for Erectile Dysfunction." *J. Androl.* 24: 6–9.

Eichna, L. W., C. R. Park, N. Nelson, S. M. Horvath, and E. D. Palmes. 1950. "Thermal Regulation during Acclimatization in a Hot, Dry (Desert-Type) Environment." *Am. J. Physiol.* 163: 585–97.

Exton, J. H., and C. R. Park. 1967. "Control of Gluconeogenesis. I. General Features of Gluconeogenesis in the Perfused Livers of Rats." *J. Biol. Chem.* 242: 2622–36.

Exton. J. H., and C. R. Park. 1968. "Control of Gluconeogenesis in Liver. II. Effects of Glucagon, Catecholamines and Adenosine 3',5'-Monophosphate on Gluconeogenesis in the Perfused Rat Liver." *J. Biol. Chem.* 243: 4189–96.

Exton, J. H., and C. R. Park. 1969. "Control of Gluconeogenesis in Liver. III. Effects of L-Lactate, Pyruvate, Fructose, Glucagon, Epinephrine and Adenosine 3'5' Monophosphate on Gluconeogenic Intermediates in the Perfused Rat Liver." *J. Biol. Chem.* 244: 1424–33.

Henderson, M. J., H. E. Morgan, and C. R. Park. 1961. "Regulation of Glucose Uptake in Muscle. IV. The Effect of Hypophysectomy on Glucose Transport, Phosphorylation, and Insulin Sensitivity in the Isolated Perfused Heart." *J. Biol. Chem.* 236: 273–77.

Henderson, M. J., H. E. Morgan, and C. R. Park. 1961. "Regulation of Glucose Uptake in Muscle. V. The Effect of Growth Hormone on Glucose Transport in the Isolated, Perfused Rat Heart." *J. Biol. Chem.* 236: 2157–61.

Jefferson, L. S, J. H. Exton, R. E. Butcher, E. W. Sutherland, and C. R. Park. 1968. "Role of Adensine 3',5'Monophosphate in the Effects of Insulin and Anti-Insulin Serum on Liver Metabolism." *J. Biol. Chem.* 243: 1031–38.

Mallette, L. E., J. H. Exton, and C. R. Park. 1969. "Control of Gluconeogenesis from Amino Acids in the Perfused Rat Liver." *J. Biol. Chem.* 244: 5713–23.

Morgan, H. E., E. Cadenas, D. M. Regen, and C. R. Park. 1961. "Regulation of Glucose Uptake in Muscle. II. Rate-Limiting Steps and Effects of Insulin and Anoxia in Heart Muscle from Diabetic Rats." *J. Biol. Chem.* 236: 262–68.

Morgan, H. E., M. J. Henderson, D. M. Regen, and C. R. Park. 1961. "Regulation of Glucose Uptake in Muscle. II. The Effects of Insulin and Anoxia on Glucose Transport and Phosphorylation in the Isolated, Perfused Heart of Normal Rats." *J. Biol. Chem.* 236: 253–61.

Morgan, H. E., D. M. Regen, M. J. Henderson, T. K. Sawyer, and C. R. Park. 1961. "Regulation of Glucose Uptake in Muscle. VI. Effects of Hypophysectomy, Adrenalectomy, Growth Hormone, Hydrocortisone and Insulin on Glucose Transport and Phosphorylation in the Perfused Rat Heart." *J. Biol. Chem.* 236: 2162–68.

Morgan, H. E., D. M. Regen, and C. R. Park. 1964. "Identification of a Mobile Carrier-Mediated Sugar Transport System in Muscle." *J. Biol. Chem.* 239: 369–74.

Park, C. R. Personal Communication.

Park, C. R., J. Bornstein, and R. L. Post. 1955. "Effect of Insulin on Free Glucose Content of Rat Diaphragm in vitro." *Am. J. Physiol.* 182: 12–16.

Park, C. R., and L. H. Johnson. 1955. "Effect of Insulin on Transport of Glucose and Galactose into Cells of Muscle and Brain." *Am. J. Physiol.* 182: 17–23.

Park, C. R., and M. Krahl. 1949. "Effect of Pituitary Extracts upon Glucose Uptake by Diaphragms from Normal, Hypophysectomized, and Hypophysectomized-Adrenalectomized Rats." *J. Biol. Chem.* 181: 247–54.

Park, C. R., and W. B. Wood Jr. 1942. "*p*-Aminobenzoic Acid as a Metabolite Essential for Bacterial Growth." *Bull. Johns Hopkins Hosp.* 70: 19.

Pilkis, S. J., T. H. Claus, I. J. Kurland, and A. J. Lange. 1995. "6-Phosphofructo-2-Kinase/Fructose-2,6-Bisphosphatase: A Metabolic Signaling Enzyme." *Ann. Rev. Biochem.* 64: 799–835.

Pilkis, S. J., M. R. El-Maghrabi, and T. H. Claus. 1988. "Hormonal Regulation of Hepatic Gluconeogenesis and Glycolysis." *Ann. Rev. Biochem.* 57: 755–83.

Regen, D. M., W. W. Davis, H. E. Morgan, and C. R. Park. 1964. "The Regulation of Hexokinase and Phosphofructokinase Activity in Heart Muscle. Effects of Alloxan Diabetes, Growth Hormone, Cortisol and Anoxia." *J. Biol. Chem.* 239: 43–49.

Thomas, M. K., S. H. Francis, and J. D. Corbin. 1990. "Characterization of a Purified Bovine Lung cGMP-Binding cGMP Phosphodiesterase." *J. Biol. Chem.* 265: 14964–70.

Turko, I. V., S. A. Ballard, S. H. Francis, and J. D. Corbin. 1999. "Inhibition of Cyclic GMP-Binding cGMP Specific Phosphodiesterase (Type 5) by Sildenafil and Related Compounds." *Mol. Pharm.* 56: 124–30.

17

Jane Harting Park

Enthusiast for Science

Research with Sidney Velick in the Cori Laboratory

Jane Harting, who is usually called Janey, was born in St. Louis on March 25, 1925. Her grandfather came from Germany and settled in St. Louis, and he owned a general store in the German section. Her father was sent to Washington University, where he became captain of the football team and graduated as a civil engineer. Her mother was artistic and played the piano.

Janey entered Washington University on a scholarship and graduated with a BA in chemistry and biology in 1946. During her time there, she worked with the distinguished embryologist Victor Hamburger, and this ignited her interest in research. She commenced medical studies there, but after two years took a leave of absence to work with the Coris, with the intention to return. However, she found the Cori laboratory to be exciting and filled with great people. She was introduced to Gerty Cori, and, as they say, the rest is history. She worked with Sidney Velick (see chapter 9) and received her PhD in 1952. Velick's area of research was the glycolytic enzyme glyceraldehyde-3-phosphate dehydrogenase, and this became the subject of much of her research over the next years. She found that the enzyme could utilize acetylaldehyde to form acetyl phosphate, and she also studied the binding of NAD by the enzyme. She thought her thesis project was rather dull, so she began a project of her own during late-night hours, working on the steps involved in the production of high-energy phosphate by the enzyme. She presented her findings on the mechanism of action of the enzyme, with its enzyme-substrate intermediate, at a meeting of the American Society of Biological Chemists in Atlantic City. Her findings contradicted the work of the great Otto Warburg. Her future husband, Charles (Rollo) Park, described her presentation in the following terms. "I was there to witness an extraordinary performance which was still spoken of years later. The science was very good but was helped by the very good looking, athletic blonde presenter and her enthusiasm." While in the Cori laboratory, she said she noticed a tall fellow (Rollo Park) at the other end, and a friend said he was a nice guy. Another friend suggested that she should date him, but she still felt that her career was too important at this stage. However, they continued to see each other during their midnight research and in the cafeteria.

On graduation, Janey worked for a time with Britton Chance at the Johnson Foundation of the University of Pennsylvania. Biographical details of this charismatic scientist are discussed in chapter 12 on Mildred Cohn. She continued to interact with Chance intermittently, studying the complex between glyceraldehyde-3-phosphate dehydrogenase and NAD. She was then accepted for postdoctoral study by the Nobel Laureate, Severo Ochoa, at New York University. While she was there, Feodor Lynen visited from Munich and suggested that she work with Fritz Lipmann, which she later did.

Marriage to Charles Park and Recruitment to Vanderbilt

With such credentials, Janey was a highly desirable faculty member, and Rollo felt the only hope of attracting her to Vanderbilt was to offer to marry her (Figure 19). In the summer of 1952, Janey departed for New York, and Rollo left for Nashville. He stayed in a rather shabby boarding house run by a member of a wealthy and prominent Nashville family, but managed to see Janey off and on. In the summer of 1953, he picked Janey up in New York and went to visit a friend in Baltimore, Victor Najjar (see chapter 10), a professor of pediatrics at Johns Hopkins whom he had known in the Cori laboratory. During that visit, he proposed to Janey, and they decided to get married within a few weeks. Neither parent was available for the wedding, Rollo's parents being ensconced at their holiday home in Cape Breton Island in Nova Scotia. So the only relative present for their marriage was a distant cousin, Henry Meade, and his wife. Janey's bridesmaid was Dr. Lillian Recant, a friend from the Cori days, and the best man was Victor Najjar. Janey wore a green designer dress, and Rollo wore a white suit. The Episcopal service proceeded without incident except that Janey was late. The wedding reception was small and simple and was hosted by the Najjar family.

The wedding couple set off with an aluminum canoe on the roof of the car and spent the night in an old inn near Belair, Maryland, that was well known for hosting newlyweds. The next day, they continued on to New York, but worried about a grinding noise in the wheels. The noise stopped whenever they stopped, and they discovered that some pranksters at the wedding had put stones in the hubcaps. In New York, they were joined by a third party, Dr. Professor Carl Martius, a stiff, formal, and distinguished scientist from the famous ETH (Swiss Federal Institute of Technology) in Zurich. Martius was a friend of Janey's but probably felt out of place as they drove to the annual meeting of the American Physiological Society in Montreal. On arrival, Janey collapsed and was bedridden for several days with a throat infection, while Rollo delivered a paper. After Montreal, they drove to Woods Hole, Massachusetts, to spend several days at the Marine Biological Station and also to canoe and fish.

Back in Nashville, they quickly found a two-room apartment over a garage. They had only the bare essentials in furniture and sat on small barrels in place of chairs. They had no bed, but slept in comfort on a pile of four mattresses sent by Rollo's parents. One of their first guests was Bernard Davis from Harvard, who was the Millikan lecturer. Davis cooked potato pancakes to go with a thick steak. They served Jack Daniels whiskey, which Janey poured down the sink at the end of the party—she was unaccustomed to alcohol and thought it wouldn't keep! After a few weeks in Nashville, Janey went to Boston to spend a year working with Fritz Lipmann at the Massachusetts General Hospital in Boston studying the mechanism by which thyroid hormones uncouple oxidative phosphorylation.

Studies on Glyceraldehyde-3-Phosphate Dehydrogenase

Janey was negatively impressed by the Physiology Department at Vanderbilt and felt they wouldn't stay long. The state of the department is described at length in chapter 16 on Rollo Park. Eventually, they received more space, equipment, and funds, and the department received national and international recognition. Janey continued her work on glyceraldehyde-3-phosphate dehydrogenase and became a leader in the field. She worked with Dan Koshland Jr. at the Brookhaven National Laboratory in Long Island to analyze the hydrolytic activity of the enzyme and determined the reactions involved in this activity and its inhibition by NAD. Koshland was an accomplished biochemist who was independently wealthy because his father was president of Levi Strauss & Co., famous for its jeans. Koshland later became editor of *Science* magazine. The enzyme proved to be fascinating, with multiple catalytic activities. For example, in addition to oxidation and phosphorylation, the enzyme was shown to catalyze transacylase, phosphatase, and esterase reactions. Furthermore, the enzyme was shown to undergo *S*- and *N*-acetylations by two model substrates—acetyl phosphate and *p*-nitrophenyl acetate, which acetylated a specific cysteine and a specific lysine respectively. Janey spent some time with Britton Chance in Philadelphia utilizing spectroscopic techniques to study the binding of NAD to the ezyme, and from these and other data, they developed a general proposal for the catalytic mechanism of the enzyme and for the effects of NAD.

In contrast to other dehydrogenases, glyceraldehyde-3-phosphate dehydrogenase was found not to require zinc for activity. However, a critical histidine residue was found in the active site, and its role in the various catalytic activities was explored by photooxidation of this residue. This work involved Judy Bond, a postdoctoral fellow, and Sharron Francis, a graduate student, and comprised much of the latter's dissertation. The career of Sharron Francis is described in the preceding chapter, and that of Judy Bond is described at the end of this chapter. In further work on the enzyme, the effects of various adenine nucleotides were examined, and it was shown that they inhibited the hydrolysis of the S-acetyl-enzyme intermediate. Further work indicated that ATP inhibited almost every step in the mechanism of this multifunctional enzyme, and evidence was presented that the effect of the nucleotide was physiological and was related to muscle function.

Research on Muscular Dystrophy and Use of EPR and NMR

In 1975, Janey made a transition to study the disease of muscular dystrophy using a hereditary avian (chicken) model of the disease and discovered that the condition could be ameliorated by penicillamine a chelating and sulfhydryl reagent. The next phase of her work involved a large series of experiments utilizing electron paramagnetic resonance (EPR). These involved some very able postdoctoral students who developed programs to analyze the data and synthesized the spin-labeled compounds used in the stuies. An important collaborator who participated was Larry Dalton, a brilliant spectroscopist from the Chemistry Department at Vanderbilt. One of the first studies showed that erythrocyte membranes from patients with Duchenne's muscular dystrophy differed from those of normal controls. An important development was the synthesis of ^{15}N- and ^{2}H-substituted maleimide spin labels for use in biological EPR studies. These spin-labeled compounds were used to investigate structure-function relationships of glyceraldehyde-3-phosphate dehydrogenase and its binding to the band-3 protein of erythrocytes. EPR studies using NAD with spin labels in the N^6 or C-8 of the adenine were combined with X-ray crystallographic data to explore the binding of NAD to the tetrameric form of the enzyme. The data showed that when one of the four NAD sites was occupied, there was no preferential binding of NAD to the other three sites. Interestingly, the solubilized enzymes from muscle and erythrocytes showed a high degree of conservation of the quaternary structure seen in the crystallized enzyme.

In 1987, Janey and Rollo took a sabbatical at the University of Pennsylvania to collaborate with Mildred Cohn and Britton Chance. They began studies utilizing nuclear magnetic resonance (NMR) to explore the changes in ATP, P-creatine, and inorganic phosphate (P_i) in exercising muscle. The project utilized ^{31}P NMR to estimate changes in the concentrations of these compounds in glycolytic and oxidative fibers in wrist flexor muscles during different levels of exercise in untrained men. The results showed that low levels of exercise primarily involved oxidative fibers and there was little lactate production. At higher work loads, glycolytic fibers were recruited and lactate production was higher. The studies were extended to include elite long-distance runners. These runners were found to have higher levels of ATP and P-creatine in their flexor muscles at rest or during exercise compared with sedentary controls. The athletes exhibited a higher force output and a lower rise in the ratio of P_i to P-creatine during exercise. Their muscles also did not lose adenine nucleotides or total phosphate, whereas the controls showed significant losses. From all of this, the Parks and their colleagues concluded that the muscles of world-class athletes had a higher capacity for generating ATP by oxidative metabolism than did control muscles.

NMR and its companion, magnetic resonance imaging (MRI), not only provide a means to explore physiological processes in a noninvasive way, as shown above, but also are now used extensively for diagnosing a wide variety of diseases and monitoring their progress. In her most recent work, Janey has utilized ^{31}P-NMR and MRI to evaluate several muscle diseases in humans and to follow their progress and determine the effectiveness of various treatments. Much of the effort has been directed to dermatomyositis, and juvenile subjects with this disease were shown to have abnormalities in Mg^{2+} and ATP levels and impaired metabolic recovery and decreased oxidative capacity. Other muscle diseases studied included fibromyalgia, scleroderma, polymyositis, and the myopathy induced by alcohol. A critical participant in the studies was Nancy Olson, a local neurologist.

Throughout her career, Janey achieved leadership roles in major scientific societies (the American Society of Biological Chemists, where she served as treasurer and member of the Publications Committee; the Federation of American Societies for Experimental Biology, where she was on the Finance Committee). She also served in several important capacities at NIH (member of the Physiological Chemistry Study Section and member (chairman) of the Board of Scientific Counselors of the National Institute of Aging and of the National Institute of Heart, Lung and Blood Diseases). Her achievements were recognized locally when she was selected as Nashville Woman oof the Year for her work on muscular dystrophy.

Throughout all these national activities, Janey maintained an active laboratory with graduate and postdoctoral students. One of her postdoctoral students, Judith Bond, went on to become an expert in metalloproteinase and to have a major career in biochemistry (chairman of the Biochemistry Department at Pennsylvania State Medical School and president of the American Society for Biochemistry and Molecular Biology).

References

Beth, A. H., K. Balsubramanian, R. T. Wilder, S. D. Venkataramu, B. H. Robinson, L. R. Dalton, C. R. Park, and J. H. Park. 1981. "Structural and Motional Changes in Glyceraldehyde-3-Phosphate Dehydrogenase upon Binding to Band 3 Protein of the Red Cell Membrane Examined with [^{15}N, ^{2}H] Maleimide Spin Label and EPR." *Proc. Natl. Acad. Sci. U.S.A.* 78: 4955–59.

Beth, A. H., B. H. Robinson, C. E. Cobb, L. R. Dalton, W. Trommer, J. J. Birktoft, and J. H. Park. 1984. "Spin-Labeled NAD^+ Bound to Glyceraldehyde-3-Phosphate Dehydrogenase: Comparison of EPR and X-ray Modeling Data." *J. Biol. Chem.* 259: 9717–28.

Chance, B., and J. H. Park. 1967. "The Properties and Enzymatic Significance of the Enzyme-Diphosphopyridine Nucleotide Compound on 3-Phosphoglyceraldehyde Dehydrogenase." *J. Biol. Chem.* 242: 5093–5105.

Chou, T.-H., E. J. Hill, E. Bartle, K. Woolley, V. LeQuire, W. Olson, R. Roelofs, and J. H. Park. 1975. "Effects of Penicillamine Treatment on Hereditary Avian Muscular Dystrophy." *J. Clin. Invest.* 56: 842–49.

Francis, S. H., B. P. Meriwether, and J. H. Park. 1971. "Interaction between Adenine Nucleotides and 3-Phosphoglyceraldehyde Dehydrogenase. I. Inhibition of Hydrolysis of S-Acetylenzyme in the Esterase Activity." *J. Biol. Chem.* 246: 5427–32.

Francis, S. H. B. P. Meriwether, and J. H. Park. 1971. "Interaction between Adenine Nucleotides and 3-Phosphoglyceraldehyde Dehydrogenase. II. A Studyof the Mechanism of Catalysis and Metabolic Control of the Multifunctional Enzme. *J. Biol. Chem.* 246: 5433–41.

Harting, J., and S. Velick. 1954. "Acetyl Phosphate Formation Catalyzed by Glyceraldehyde-3-Phosphate Dehydrogenase." *J. Biol. Chem.* 207: 857–65.

Harvey, M. S., J. H. Park, and F. Lipmann. 1955. "Magnesium Antagonism of the Uncoupling of Oxidative Phosphorylation by Iodothyronines." *Proc. Natl. Acad. Sci. U.S.A.* 41: 571–76.

Niermann, K. J., J. H. Park, R. J. Meyer, D. A. Gilpin, L. E. King, and N. J. Olsen. 2001. "Comparison of Magnesium (Mg^{2+}) and ATP Abnormalities in Muscle Disorders; Fibromyalgia, Dermatomyositis and Scleroderma." *Int. Soc. Magnetic Resonance Med.* 9: 1773.

Oguchi, M., B. P. Meriwether, and J. H. Park. 1973. "Interaction between Adenosine Triphosphate and Glyceraldehyde-3-Phosphate Dehydrogenase. III. Mechanism of Action and Metabolic Control of the Enzyme under Simulated in vivo Conditions." *J. Biol. Chem.* 248: 5562–70.

Olson E. J., and J. H. Park. 1964. "Studies on the Mechanism and Active Site for the Esterolytic Activity of 3-Phosphoglyceraldehyde Dehydrogenase." *J. Biol. Chem.* 239: 2316–26.

Park, C.R. Personal Communication.

Park, J. H. Personal Communication.

Park, J. H., C. F. Agnello, and J. H. Park. 1966. "S-N Transfer and Dual Acetylation in theS-Acetylation and N-Acetylation of 3-Phosphoglyceraldehyde Dehydrogenase." *J. Biol. Chem.* 241: 769–71.

Park, J. H., R. L. Brown, C. R. Park, M. Cohn, and B. Chance. 1988. "Energy Metabolismof the Untrained Muscle of Elite Runners as Observed by ^{31}P-Magnetic Resonance Spectroscopy: Evidence Suggesting a Genetic Endowment for Endurance Exercise." *Proc. Natl. Acad. Sci. U.S.A.* 85: 8780–84.

Park, J. H., R. L. Brown, C. R. Park, K. McCully, M. Cohn, J. Haselgrove, and B. Chance. 1987. "Fractional Pools of Oxidative and Glycolytic Fibers in Human Muscle Observed by ^{31}P-Magnetic Resonance Spectroscopy during Exercise." *Proc. Natl. Acad. Sci. U.S.A.* 84: 8976–80.

Park, J. H., and D. E. Koshland Jr. 1958. "The Hydrolytic Activity of Glyceraldehyde-3-Phosphate Dehydrogenase." *J. Biol. Chem.* 233: 986–90.

Park, J. H., B. P. Meriwether, P. Clodfelder, and L. W. Cunningham. 1961. "The Hydrolysis of *p*-Nitrophenyl Acetate Catalyzed by 3-Phosphoglyceraldehyde Dehydrogenase." *J. Biol. Chem.* 236: 136–41.

Taylor, E. L., B. P. Meriwether, and J. H. Park. 1963. "The Hydrolysis of *p*-Nitrphenyl Acetate Catalyzed by 3-Phosphoglyceraldehyde Dehydrogenase from Yeast." *J. Biol. Chem.* 238: 734–40.

Velick, S., J. E. Hayes, and J. Harting. 1953. "The Binding of Diphosphopyridine Nucleotide by Glyceraldehyde-3-Phosphate Dehydrogenase." *J. Biol. Chem.* 203: 527–44.

Wilkerson, L. S., R. C. Perkins Jr., R. Roelofs, L. Swift, L. R. Dalton, and J. H. Park. 1978. "Eryhrocyte abnormalities in Duchenne Dystrophy Monitored by Saturation Transfer Electron Paramagnetic Resonance Spectroscopy." *Proc. Natl. Acad. Sci. U.SA.* 75: 838–41.

18

Gerty Cori's Work on Glycogen Structure and Glycogen Storage Diseases

Humans suffer from several types of glycogen storage diseases, which mainly affect the liver and are marked by enlargement of this organ due to glycogen accumulation and, in some cases, hypoglycemia. There are other glycogen storage diseases in muscle where deficiency of enzymes involved in carbohydrate metabolism leads to weakness and cramping. To elucidate the nature of these diseases, Gerty Cori first turned her attention to glycogen itself, which is a very large molecule with an interesting tree-like structure with many branches. The linear portions are made up of 1,4-linkages between glucose residues, but the branch points involve 1,6-linkages. Earlier work in the laboratory had shown that highly purified phosphorylase, which is specific for 1,4-linkages, could only partially degrade glycogen, whereas crude phosphorylase could hydrolyze it completely. As described in detail in chapter 19 devoted to Joseph Larner, it was shown that a second enzyme, initially called debranching enzyme and later amylo 1,6-glucosidase, needed to work with phosphorylase to completely degrade glycogen. In other studies in which a postdoctoral student Barbara Illingworth (see chapter 20) was involved, these enzymes and some chemical methods were then used to determine the structure of glycogen—the degree of branching and the number of glucose residues in the inner and outer branches.

Gerty's main interest in her later years was to define the enzyme deficiencies underlying certain glycogen storage diseases. In her laboratory stood a cabinet with samples of glycogen that she had isolated from liver and other tissues sent to her by numerous clinicians. Larner correctly proposed that one of these diseases, a form of Von Gierke's disease, was attributable to an abnormality of glycogen structure due to a lack of the debranching enzyme. However, Gerty bet that it was due to the absence of glucose-6-phosphatase, the enzyme that catalyzes the last step in the formation of glucose from glycogen. Who won the bet is described in the next chapter.

The identification of the enzymes involved in glycogen storage diseases was of significance because only one other molecular disease, sickle cell anemia, was known at that time. Gerty characterized other forms of von Gierke's disease. Two fatal cases showed a total lack of glucose-6-phosphatase in the liver, and other milder cases showed a lower than normal level. In another study with Illingworth, a generalized form of glycogen storage disease with altered glycogen structure was found to be due to a deficiency of amylo 1,6-glucosidase.

Gerty's Personality and Philosophy

Larner describes Gerty as a "tireless scientific worker and an avid reader. She was at all times a superb experimentalist and analyst with the most demandingly high standards." She personally taught him how to perform simple as well as complex laboratory procedures. Larner states that Gerty had "vivacity and a love of science and discovery that were infectious." She smoked incessantly, and cigarette ash was dropped constantly. Like Carl, she had an instinctive feel for the right experiment, but she would often have spontaneous thoughts that were extremely original. This was in contrast to Carl, who was more reflective and would design more rigorous, decisive experiments. Again, Larner informs us that "she needed only one exciting experimental finding to jump into a problem with unbounded energy."

Gerty resembled Carl in her love of books. She would order five to seven books per week from a local library, and these would be delivered either to the laboratory or the department office. She would read them all in a week. She had the same intellectual breadth as Carl, being conversant with political theory, sociology, and art and the humanities. As Gerty's health slowly declined, she had to spend less time in the laboratory and more time resting on a cot in her office. She worked almost to the very end, despite the increasing anemia, swelling of the legs, and enlargement of the spleen and liver. A splenectomy was performed, but it did not stop the relentless progress of the disease. Carl would carry her everywhere when she was too weak to stand, and when she was too sick to come to the department, he would care for her at home. When she died on October 26, 1957, Carl left for a month and walked the beaches of the Caribbean, looking much healthier when he returned. Eighteen months later he met Anne Fitzgerald, who became his second wife. She was a friend of the family who had lost her husband.

Gerty wrote a short statement of her personal philosophy for the National Academy of Sciences in 1954. It was titled "This I Believe" and describes her personal beliefs. Her statement was included in a series of radio programs by Edward R. Murrow, who was famous for his broadcasts from London during World War II. Gerty first recognized her gratitude for how she and Carl were treated in the United States. She then described the benefits of living in two cultures and expressed her belief that art and science are both glories of the human mind.

Although she had left behind the utopian hopes of her youth, she still had hope about the future despite the terrible wars of the century. She believed that the greatest achievements in art and science came from men who had faith or compassion for their fellow men. She decried totalitarianism and the fact that science had given tools of great effectiveness to ruthless men with an excessive will to dominate their fellow men. She felt that her beliefs had undergone little change in her life, and she liked to think that they had developed into a somewhat higher plane. Honesty, intellectual integrity, courage, and kindness were still the virtues she admired, although with advancing years, kindness seemed more important. She considered that the love for and dedication to one's work seemed to be the basis for happiness. For a research worker, the unforgotten moments are those rare ones, which come from years of plodding work, when the veil over nature's secrets seems suddenly to lift, and when what was dark and chaotic appears in a clear and beautiful light and pattern.

In addition to the Nobel Prize for which she was the first U.S. female recipient and second woman overall after Marie Curie, Gerty was elected to the National Academy of Sciences and received an award from the American Chemical Society, the Squibb Award from the Endocrine Society, the Garven Medal of the American Chemical Society, the St. Louis Award, the Borden Award of the Association of American Medical Colleges, and many honorary degrees. At various times, Carl commented about his collaboration with Gerty: "It is a delicate operation which requires much give and take on both sides and occasionally leads to friction, if both are equal partners and not willing to yield on a given point," and: "Our efforts have been largely complementary, and one without the other would not have gone so far as in combination." In 2008, the U.S. Postal Service issued a stamp honoring Gerty's accomplishments—with an incorrect structure of glucose-1-phosphate.

In a memorial service for Gerty at Washington University, Bernardo Houssay spoke the following words:

> Gerty Cori's life was a noble example of a dedication to an ideal, to the advancement of science for the benefit of humanity. Her charming personality, so rich in human qualities, won the friendship and admiration of all who had the privilege of knowing her. Her name is engraved forever in the annals of science, and her memory will be cherished by all her many friends as long as we live.

References

Cori, C. F. 1969. "The Call of Science." *Ann. Rev. Biochem.* 38: 1–21.
Cori, G. T. 1952. "Glycogen Structure and Enzyme Deficiencies in Glycogen Storage Disease." *Harvey Lect.* 48: 145–71.

Cori, G. T., and C. F. Cori. 1952. "Glucse-6-Phosphatase of the Liver in Glycogen Storage Disease." *J. Biol. Chem.* 199: 661–67.

Cori, G. T., and J. Larner. 1951. "Action of Amylo-1,6-Glucosidase and Phosphorylase on Glycogen and Amylopectin." *J. Biol. Chem.* 188: 17–29.

Illingworth, B., G. T. Cori, and C. F. Cori. 1956. "Amylo-1,6-Glucosidase in Muscle Tissue in Generalized Glycogen Storage Disease." *J. Biol. Chem.* 218: 123–29.

Illingworth, B., J. Larner, and G. T. Cori. 1952. "Structure of Glycogens and Amylopectins. I. Enzymatic Determination of Chain Length." *J. Biol. Chem.* 188: 631–40.

Larner, J. 1992. "Gerty Theresa Cori 1896–1957." *Biogr. Mem. Natl. Acad. Sci. U.S.A.* 61: 110–35.

Larner, J., B. Illingworth, G. T. Cori, and C. F. Cori. 1952. "Structure of Glycogens and Amylopectins. II. Analysis by Stepwise Enzymatic Degradation." *J. Biol. Chem.* 199: 641–51.

19

Joseph Larner

Focus on Glycogen Synthase

Early Life and Research in the Cori Laboratory

When Joseph Larner's parents arrived in the United States from Brest-Litovsk in 1921, he was a ten-month-old baby. Brest-Litovsk was the site of a separate peace treaty between Soviet Russia and Germany in 1918 toward the end of World War I. The family name was originally Likovsky, and the children changed it to Larner at various stages in their careers. The family settled first in New York City, where his father worked in a furniture factory. His parents then moved to Newark, New Jersey, where they had a kosher butcher shop, and later to Orange, New Jersey, where they had a candy and general small item store. It was a seven-day, eighteen-hours-a-day operation, and Joe Larner and his younger brother helped out by opening the store, sorting and delivering the newspapers, and stoking the coal furnace. They were then relieved by their mother and given breakfast before leaving for school. Larner and his younger brother and sister attended public schools in Orange before attending the University of Michigan. When war came in 1941, he was unable to join the military because of severe nearsightedness. He instead entered Columbia College of Physicians and Surgeons, and, since the medical course was compressed to three years, he graduated in 1945. When the medical requirements for physicians were lowered, he entered the army, and, upon graduation, finished as a first lieutenant. At Columbia, DeWitt Stetten—who investigated uric acid metabolism and would later achieve fame at NIH holding several offices and serving as Director of the National Institute of General Medical Sciences—was Larner's biochemistry teacher. When it became time for Larner to choose a laboratory for doctoral work, Stetten introduced him, while he was still in uniform, to Carl Cori on an elevator at a FASEB meeting in Chicago. After an abbreviated internship at the University of Chicago, Larner completed his military service at the Edgewood Arsenal in Maryland, working on a kit to detect nerve gas in the field. His supervisor was

Bill Summerson, the co-inventor of the Klett-Summerson photometer. Larner's research involved studying the effects of warfare agents on pyruvate dehydrogenase, and this work resulted in a publication in the *Journal of Biological Chemistry* showing the enzyme's requirement for NAD.

Carl Cori accepted Larner into the graduate program in his department, but felt that he first needed a master's degree in chemistry. Since most of the chemists at Washington University were working on the atom bomb project at Los Alamos, he was advised to go to the University of Illinois. The chairman there was William C. Rose, who discovered the amino acid threonine and whose interaction with du Vigneaud is described in chapter 12 on Mildred Cohn. Larner describes Rose as a master teacher and a gentleman of the "old Southern school." Another influential teacher was Herbert Carter, who lectured on vitamins and hormones and who introduced Larner to the polyol inositol. At that time, its biological role was unknown, and Larner became intrigued with finding a function for it.

On arrival in St. Louis, Larner was assigned to Gerty Cori, and the first thing she required him to do was to demonstrate his pipetting skills while she looked over his shoulder! When he passed that test, he was allowed to do real research. His project was to recrystallize phosphorylase ten times and to confirm that it only partially (40 percent) degraded glycogen, whereas the crude enzyme degraded it completely. As described earlier, the Coris had found that the products of phosphorylase action on glycogen were glucose-1-phosphate, the Cori ester, but, unexpectedly, also free glucose. The explanation was that the phosphorylase was contaminated by another enzyme even though it was repeatedly recrystallized. This idea was supported by the work of Shlomo Hestrin, who had preceded Larner in the laboratory. Larner's first project was to purify the contaminating enzyme, which was postulated to release glucose from the branch points of the glycogen. This enzyme was designated debranching enzyme or amylo-1,6-glucosidase, and, when it was tested on glycogen, it was shown to release free glucose. Since other explanations were possible, Gerty was surprised and exhilarated by the result. Larner describes how she ran from the laboratory to Carl's office shouting, "Carly, Carly, it's free glucose, it's free glucose!"

Larner continued to work with Gerty to determine the degree of branching in liver and muscle glycogen and amylopectin (a form of starch) using the combined enzymes and measuring the ratio of free glucose to glucose-1-phosphate. In collaboration with Barbara Illingworth, the enzymes were used to determine the number of glucose residues in the outer and inner branches of samples of liver and muscle glycogen obtained from different species. In a final study, which completed Larner's dissertation for his PhD, he used phosphorylase and amylo-1,6-glucosidase in a sequential manner to find out the arrangement of the branch points in glycogen and amylopectin. The results were consistent with only one model, namely, that both polysaccharides had a multibranched, tree-like structure. Larner extended this work by looking at the action of branching enzyme on glycogen, which contained

1,4-linked^{14}C glucose in the outer chains. These experiments showed that the enzyme was an amylo-(1,4 → 1,6)-transglucosidase.

Larner's next project was to work with Gerty on elucidating the enzymatic defect in a form of Von Gierke's disease. Larner thought it might be due to a deficiency in amylo-1,6-glucosidase, but Gerty thought it was due to a lack of glucose-6-phosphatase. If he was right, the glycogen should have an abnormal structure with shortened outer branches, whereas if she were right, the glycogen would be normal. They made a bet on the outcome, but when the patient's glycogen was examined by iodine staining, a very abnormal result was obtained. Instead of giving the usual brown stain, it gave a purple color like starch. They concluded that the glycogen structure was abnormal because the outer branches had been built up because of good nutrition. Gerty then undertook to elucidate the enzymatic basis for other forms of glycogen storage disease.

While in the Cori laboratory, Larner interacted with Earl Sutherland, whose laboratory was across from his and with whom he became very good friends. They would have many subsequent interactions. Another person with whom Larner interacted was William Daughaday, whose career will be described later (chapter 21). Daughaday was interested in the urinary excretion of *myo*-inositol in humans and animals with diabetes. This was later shown not to be due to the diabetes per se but to an inhibitory effect of the high concentration of glucose on the renal tubular uptake of *myo*-inositol. Another project involved the demonstration that glucose could be converted to *myo*-inositol in embryonic tissues. Larner particularly consulted Oliver Lowry, who was chairman of the Department of Pharmacology and a superb methodologist. Larner was impressed with the fact that Lowry continued work in the laboratory with his own hands.

To complete his PhD, Larner had to demonstrate proficiency in a foreign language. He failed his first attempt on the exam in German, which dealt with old-fashioned names for minerals in spas. Gerty offered to tutor him for the second attempt, and he passed. An oral exam was also required, and Carl's first question was: "Tell me everything you know about glutathione." Larner started with Sir Frederick Gowland Hopkins, who had won the Nobel Prize in 1929 for his work on vitamins and had determined the structure of glutathione. After several hours of discussion, Larner was duly granted the degree.

Move to Western Reserve University

At the end of his stay in the Cori laboratory, Larner accepted a position in biochemistry at the University of Illinois and began working on the digestion of starch in the gastrointestinal tract and the role of glycosidases. He also began a study of the effects of insulin on glycogen synthesis and on phosphorylase in rat diaphragms. However, no effect on phosphorylase could be detected. Larner was not happy with the large amount of teaching he was expected to do (150 hours per year), and when Earl Sutherland, who was now chairman of pharmacology at Western Reserve University, sent him a letter inquiring if he was interested in a position there, Larner jumped at the opportunity. He was appointed associate professor and housed in a renovated laboratory in the old medical school building with funds provided for diabetes research.

He spent seven years in Cleveland, and his first postdoctoral hire was Carlos Villar-Palasi, who came from Spain and became a long-term collaborator. Sutherland wanted someone in the department to work on insulin and suggested that Larner compare the effects of insulin stimulation of glycogen synthesis with the effects of high glucose in rat diaphragms. This involved measurements of glucose-1-phosphate and glucose 6-phosphate. He found that insulin produced a large increase in glycogen synthesis and an increase of glucose-6-phosphate, but he found no change in glucose-1-phosphate. These results suggested an effect of insulin on glycogen synthase but not on phosphorylase. As described earlier, Luis Leloir had discovered glycogen synthase and had shown that it, rather than phosphorylase, was responsible for glycogen synthesis. Larner and Villar-Palasi had learned via the "grapevine" that glycogen synthase could be activated by glucose-6-phosphate. Thus the increase in glucose-6-phosphate induced by insulin (due to its effect on glucose uptake in the diaphragms) could explain the increased activity of glycogen synthase. Since Leloir had shown that UDPG was the substrate of the synthase, some mechanism had to be found to generate UDPG. This entailed the enzyme UDPG pyrophosphorylase, which synthesizes UDPG from UTP and glucose-1-phosphate with pyrophosphate as the other product, and glucose-1-phosphate being generated from glucose-6-phosphate through the action of phosphoglucomutase. However, although this sequence of metabolic events could explain how insulin increased glycogen synthase activity, further work revealed a problem. This was that the activation of the enzyme by insulin in rat diaphragms was not lost when glucose-6-phosphate was removed from the extracts by dialysis and ammonium sulfate precipitation of the enzyme. In other words, insulin had converted the enzyme to a stable form that did not need glucose-6-phosphate for activity.

Hormonal Regulation of Glycogen Synthase

As noted above, insulin could convert glycogen synthase to a form that had increased activity in the absence of glucose-6-phosphate. This stable form of the enzyme with increased activity was termed "independent," whereas the activity seen in the presence of excess glucose-6-phosphate was termed "total." Follow-up studies in association with Maria Rosell-Perez, a postdoctoral fellow from Barcelona, demonstrated that two forms of the enzyme could be purified that differed in their response to glucose-6-phosphate, namely, a glucose-6-phosphate independent (I) form and a glucose-6-phosphate dependent (D) form. These findings had similarity to the active and inactive forms of phosphorylase, and although Villar-Palasi was dubious, a graduate student, Daniel Friedman, set about testing to see if the D and I forms could be interconverted in muscle extracts. He found that addition of ATP and Mg^{2+} caused a conversion of the I to D form and, most importantly, using ^{32}P-labeled ATP, he showed that the conversion was associated with the incorporation of ^{32}P into the enzyme. Furthermore, when the labeled enzyme was added to a crude homogenate, it was converted to the I form and inorganic ^{32}P was released. There was a complete correspondence between the D to I conversion and the release of phosphate. These results showed that the interconversion of the two forms of the enzyme was associated with a change in their phosphorylation state. Since the reactions appeared to be enzymatic in nature, the I to D conversion was postulated to involve a protein kinase and the D to I conversion to a protein phosphatase. Thus the system was analogous to that regulating phosphorylase, except for the fact that the phosphorylated enzyme was less active than the dephosphorylated form.

The different forms of the enzyme were purified and characterized chemically and kinetically and were shown to undergo the same interconversion as seen in the cruder tissue preparations. It was also shown that the interconversion occurred in muscle of other species and in liver. Importantly, Rosell-Perez and Larner showed that addition of cyclic AMP to the kinase reaction stimulated the conversion of the I to the D form of the muscle enzyme. The researchers noted the similarity to the phosphorylase kinase system that activates muscle phosphorylase, and Larner tested phosphorylase kinase to see if it had an effect on the I to D conversion, but neither the inactive nor the active form of the kinase had any effect. Since they knew the group of Krebs and Fischer had discovered that phosphorylase kinase could be activated by phosphorylation and that this could be stimulated by cyclic AMP (see chapter 11 devoted to Krebs), they felt that a similar mechanism could operate with glycogen synthase. In retrospect, it appears that the protein kinase—discovered by Friedman and Larner in 1963 and shown by Rosell-Perez and Larner in 1964 to be stimulated by cyclic AMP—was the same as the protein kinase (PKA) discovered by Krebs and Fischer in 1968 as the activator of phosphorylase kinase. However, its overall role as a mediator of hormone and cyclic AMP effects was not recognized at that time. In collaboration with Fred Sanger of the Medical Research Council Laboratory of Molecular Biology in Cambridge, United Kingdom, Larner showed that the amino acid sequence at the site of phosphorylation of glycogen synthase was the same as that at the phosphorylation site of phosphorylase kinase, supporting the idea that the same protein kinase was involved.

In 1963–1964, Larner spent a sabbatical in the Biochemistry Department at the University of Minnesota. During that time, some of the preceding findings were reported, and a further exploration of the role of cyclic AMP in the regulation of glycogen synthase was undertaken. An earlier report by Enrique Belocopitow from the Fundación Campomar in Buenos Aires had shown that epinephrine treatment of rat diaphragms decreased glycogen synthase activity measured in the presence of glucose-6-phosphate. This was confirmed by Larner and James Craig, who also showed that a low concentration of epinephrine decreased the *I* form, whereas insulin increased it. As expected, epinephrine activated phosphorylase, but insulin had no effect on this enzyme. Although the mechanism of the effect of epinephrine to convert the *I* to the *D* form of glycogen synthase could now be explained as being attributable to the phosphorylation and inactivation of the enzyme by PKA, the mechanism of the insulin effect remained unclear and controversial. Earl Sutherland was convinced that it was due to a decrease in cyclic AMP, but this could not be demonstrated except when the nucleotide was elevated (e.g., by epinephrine).

Larner and Sutherland had quite a few interchanges about this topic, and Sutherland remained convinced that insulin acted on glycogen synthase by lowering cyclic AMP since it seemed very logical to him. Larner devoted an entire article to this topic in *Metabolism* and pointed out that it was well established that insulin caused the conversion of the *D* form of the synthase to the *I* form without a change in cyclic AMP, and that insulin acted by decreasing the sensitivity of the synthase kinase (now recognized as PKA) to cyclic AMP.

At the University of Minnesota, Larner worked with Fred Huijing, on leave from the University of Amsterdam, to study the activation of synthase kinase by cyclic AMP. He also worked with Jonathan Bishop to study the in vivo effects of insulin and glucagon on glycogen synthase in liver. They observed effects on the interconversion of the *D* and *I* forms that were similar to those for insulin and epinephrine in muscle. Back in Cleveland, further experiments using rat diaphragms confirmed the effects of insulin and epinephrine on the two forms of the enzyme and showed that insulin produced no detectable effect on the level of cyclic AMP or the activity of phosphorylase.

Move to the University of Virginia and Recruitment of Alfred Gilman

In 1969, Larner was appointed to the chair of pharmacology at the University of Virginia Medical School. He spent twenty-one years in that position and continues to be involved in insulin research. He recruited some stellar young scientists to the faculty, two of whom (Alfred Gilman and Ferid Murad) had trained in Earl Sutherland's department and became Nobel laureates. Gilman's father Alfred was on the faculty of the Pharmacology Department at Yale. The elder Gilman had coauthored a classic pharmacology textbook with Louis Goodman, a colleague at Yale, and named his son Alfred Goodman Gilman in honor of his collaborator. Gilman's friend, the Nobel Laureate Michael Brown, said that Gilman was probably the only person ever named after a textbook. After graduating with a BS from Yale in 1962, Gilman entered the MD, PhD program at Western Reserve, where he studied under Theodore (Ted) Rall Sutherland's collaborator in his historic studies that led to the discovery of cyclic AMP. His project was to study the role of cyclic AMP in the regulation of thyroid secretion by thyroid-stimulating hormone, a topic that did not excite him. He then did a three-year postdoctoral fellowship with the Nobel Laureate Marshall Nirenberg at NIH. His project was to measure cyclic AMP metabolism in neuronal cell lines, but his major accomplishment there was developing a simple method for measuring cyclic AMP based on its binding to PKA. In 1971, Gilman moved to Charlottesville to become an assistant professor of pharmacology in Larner's department. There, he began his studies of the regulation of adenylyl cyclase using mutants or hybrids of a lymphoma cell line developed by Gordon Tomkins at the University of California, San Francisco. Tomkins was a brilliant researcher who later died prematurely as a result of an operation to remove an acoustic neuroma. Henry Bourne produced a variant of the cells that appeared to lack adenylyl cyclase since they showed no activation of adenylyl cyclase when exposed to agonists that interact with β-adrenergic receptors. Somewhat unexpectedly, the defect in the mutants was found not to be in the receptors. In a brilliant series of experiments, Gilman and his associates showed that the defect was in a component required for the coupling of the receptors to the cyclase. This regulatory component was then shown to be the target of the stimulatory effect of sodium fluoride and guanine nucleotides on the cyclase. The examination of the effects of guanine nucleotides arose from previous work by Martin Rodbell, Thomas Pfeuffer, Svi Selinger, and their associates, who had implicated these nucleotides in the coupling of receptors to the cyclase. Gilman and his coworkers set out to extract and purify the regulatory component and to demonstrate that it could bind and be activated by a GTP analog and could reconstitute β-adrenergic agonist-stimulated adenylyl cyclase in the deficient lymphoma cell line.

Gilman's Discovery of G Proteins

Gilman's work at Charlottesville, which would later lead to his Nobel Prize, was accomplished with some superior postdoctoral fellows, including Elliott Ross and Paul Sternweis, who moved with him to the University of Texas Southwestern Medical School in Dallas when he became chairman of Pharmacology there in 1981. Gilman continued his work on the regulatory components, which were called guanine nucleotide-binding regulatory components and later simply G proteins. In addition to G_s, which stimulated adenylyl cyclase, he identified another type (G_i), which inhibited adenylyl cyclase. Both were found to have multiple subunits, first α- and β-subunits and then also γ-subunits. All these were purified and characterized, and they recognized that the α-subunit was responsible for binding the guanine nucleotides and that the β- and γ-subunits existed as a dimer. A significant accomplishment was the reconstitution of catecholamine-stimulated adenylyl cyclase using three purified proteins (β-adrenergic receptors, Gs, and adenylyl cyclase). In later work, the crystal structures of the α_i-subunit and the undissociated heterotrimeric protein were determined, with Steve Sprang playing a critical role. Virtually simultaneously, the structures of the α-subunit of transducin and its heterotrimeric complex were solved by Paul Sigler's group at Yale, and the essential features were the same as those reported by Gilman's group. A later major advance by Gilman's group was the construction of a soluble form of adenylyl cyclase that could be activated by the α-subunit of G_s. The crystal structure of this in a complex with the α-subunit was later determined.

For his work on G proteins, Gilman received the Nobel Prize in 1994, together with Martin Rodbell. Gilman became dean at Southwestern in 2005 and subsequently provost and vice president. He is now emeritus and in charge of a research foundation in Texas. He is a member of the National Academy of Sciences from whom he won the Richard Lounsbery Award in 1987. Gilman won the Albert Lasker Basic Medical Research Award in 1989, and the next year he won the Passano Foundation Award and the American Heart Association Basic Science Research Award. He was elected to the American Academy of Arts and Sciences in 1988 and to the Institute of Medicine of the National Academy of Sciences in 1989. In 1980, 1985, and 1990, he was the primary editor for the well-known pharmacology textbook, *The Pharmacological Basis of Therapeutics*, which was started by his father as described above. Interestingly, Gilman won the Louis S. Goodman and Alfred Gilman Award in Drug Receptor Pharmacology from the American Society of Pharmacology and Experimental Therapeutics in 1990. When Gilman retired from the University of Texas, he became the chief scientific officer of the Cancer Prevention and Research Institute of Texas, where his role was to ensure that high-quality research is supported. However, in 2012 he resigned when it was discovered that some investigators were receiving funds by bypassing the rigorous review procedures. This revelation triggered a mass resignation of the scientific advisors.

Recruitment of Ferid Murad and His Work on the Function of Nitric Oxide

Another of the important recruits Larner brought to Charlottesville was Ferid Murad. Murad's father, Jabir Murat Ejupi, was born in Albania, and when he emigrated to the United States, the officer at Ellis Island named him Murad. Ferid Murad won a scholarship to DePauw University, a small liberal arts school in Indiana, and then went to Western Reserve to enter the new MD, PhD program in 1958. He went there on the recommendation of Earl Sutherland's son Bill. Murad's mentors were Earl Sutherland and Ted Rall, who had just discovered cAMP. His project was to show that the effects of epinephrine and other catecholamines on adenylyl cyclase were mediated by β-adrenergic receptors. He also worked on the effects of cyclic AMP analogues synthesized by Theo Posternak, and this initiated his interest in cyclic GMP. To supplement his income, Murad moonlighted at the obstetric service at the Cleveland Clinic. He went to the Massachusetts General Hospital (MGH) for his internship and residency in medicine and then to NIH to work with Martha Vaughan at the Heart Institute for three years, where he studied cyclic AMP as the mediator of hormone effects in fat cells, kidney, bone, heart, and testes. He was recruited in 1970 to the University of Virginia to develop a Clinical Pharmacology Division in the Department of Medicine, with appointments in the Departments of Medicine and Pharmacology. For his initial research, Murad decided to examine the hormones and other agents that increased cyclic GMP in various tissues, but then he started to look at the regulation of guanylyl cyclase in cell-free systems. Surprisingly, certain nitrogenous compounds (azide, hydroxylamine, nitrite) were found to activate some forms of the enzyme. The stimulation with azide was enhanced by adding extracts from different tissues, and they searched for the factors that were responsible.

At that time, Murad's colleagues thought he was crazy to investigate the effects of azide, which was a known metabolic poison. However, he showed that the factor required for azide activation was catalase, and old literature had shown that the interaction of azide with catalase generated nitric oxide (NO). Other work showed that azide and related compounds increased cGMP levels in many tissues and, in the case of smooth muscle, caused relaxation. The results with catalase implicated NO as the active principal in the regulation of guanylyl cyclase by azide. The role of this gas was proved in a dramatic experiment in December 1976, which showed that chemically generated NO could activate guanylyl cyclase and increase cyclic GMP in several tissues. The idea that a toxic gas and a free radical that was an air pollutant could act like a hormone was revolutionary, and it was greeted with great skepticism. An essential requirement of the hypothesis was to show that NO could be generated in cells, and it was eight years before a sufficiently sensitive assay was developed. Along the way, Murad and his coworkers identified multiple forms of the cyclase (soluble and particulate) with only the soluble form responding to NO.

In 1981, Murad went to Palo Alto for a position as acting chairman of medicine at Stanford and then left to become a vice president at Abbott Laboratories. He continued his work on guanylyl cyclase, which encompassed an endothelial-derived relaxing factor (EDRF) that had been discovered by Robert Furchgott of the State University of New York. The similarity of its properties to vasodilators led Murad to suggest it acted by increasing cGMP. This proved to be the case, and the ability of other vasodilators to raise cGMP was also demonstrated. Atrial natriuretic peptide, which is a vasodilator and increases sodium excretion in the urine, was also found to activate the enzyme. In 1998, Murad, Furchgott, and Louis Ignarro received the Nobel Prize for their independent work on the role of NO in cell signaling. At the time of the award, Murad had taken a position at the University of Texas Houston as the first chairman of a newly combined basic science department (Integrative Biology, Pharmacology, and Physiology). In 2011, he took a position as university professor at George Washington University.

Meanwhile, at the University of Virginia, Larner continued working on the hormonal regulation of glycogen synthase in muscle, liver, and adipose tissue with several postdoctoral students, some of whom would later assume chairmanships and deanships. In particular, they demonstrated that the enzyme was multiply phosphorylated, and the relationship between its phosphorylation state and kinetic properties indicated that not all of the phosphorylation sites were equivalent. Follow-up work revealed the presence of multiple protein kinase activities in muscle that phosphorylated different sites on the synthase. Some postdoctoral fellows who were instrumental in this work were Peter Roach and his wife Anna DePaoli-Roach, Joan Guinovart, John Lawrence, Jr., and Keith Schlender.

Larner's Research on the Mechanism of Insulin Action

Larner maintained his interest in the mechanism of the action of insulin and pursued the earlier observation that the hormone inhibited the action of cyclic AMP on PKA. This observation led him to search for the messenger that mediated this action of insulin. Purification of the messenger indicated a molecular mass of less than one thousand. Independently, Leonard Jarett at Washington University had found that insulin produced a factor that activated pyruvate dehydrogenase in mitochondria. Larner obtained evidence that the two factors were identical. However, work in the laboratory of Michael Czech at the University of Massachusetts indicated that the mediator was a fragment of albumin and therefore an artifact. Around this time, the insulin receptor was shown to encode a tyrosine kinase activity, and support developed for a signaling system based on this. However, it remained unclear how this system activated glycogen synthase, and Larner continued to work on identifying the insulin messenger. He has evidence that it is a complex molecule containing *chiro*inositol and a glycan. As of this writing, he continues to work on its mechanism of action.

References

Bishop, J. S., and J. Larner. 1967. "Rapid Activation-Inactivation of Liver Uridine Diphosphate Glucose-Glycogen Transferase and Phosphorylase by Insulin and Glucagon in vivo." *J. Biol. Chem.* 242: 1354–56.

Coleman, D. E., A. M. Berghuis, E. Lee, M. E. Linder, A. G. Gilman, and S. R. Sprang. 1994. "Structures and Active Conformations of Gi Alpha 1 and the Mechanism of GTP Hydrolysis." *Science* 265: 1405.

Cori, G. T., and J. Larner. 1951. "Action of Amylo-1,6-Glucosidase and Phosphorylase on Glycogen and Amylopectin." *J. Biol. Chem.* 188: 17–29.

Craig, J. W., and J. Larner. 1964. "Influence of Epinephrine and Insulin on Uridine Diphosphate Glucose-Alpha-Glucan Transferase and Phophorylase in Muscle." *Nature* 202: 971–73.

Daughaday, W. H., and J. Larner. 1954. "The Renal Excretionof Inositol in Normal and Diabetic Human Beings." *J. Clin. Invest.* 33: 1075–80.

DePaoli-Roach, A. A., P. J. Roach, and J. Larner. "Multiple Phosphorylation of Rabbit Skeletal Muscle Glycogen Synthase. Comparison of the Actions of Different Protein Kinases Capable of Phosphorylating in vitro." 1979. *J. Biol. Chem.* 254: 12062–68.

Friedman, D. L., and J. Larner. 1963. "Studies on UDPG-Alpha-Glucan Transglucosylase.III. Interconversion of Two Forms of Muscle UDPG-Alpha-Transglucosylase by a Phosphorylation-Dephosphorylation Sequence." *Biochemistry* 2: 669–75.

Gilman, A. G. 1987. "G Proteins: Transducers of Receptor-Generated Signals." *Ann. Rev. Biochem.* 56: 615–49.

Gilman, A. G. 1994. *Les Prix Nobel. The Nobel Prizes*. Edited by Tore Frängsmyr. Stockholm: Nobel Foundation.

Huijing, F., and J. Larner. 1966. "On the Mechanism of Action of Adenosine 3',5'Cyclic Phosphate." *Proc. Natl. Acad. Sci. U.S.A.* 56: 647–53.

Illingworth, B., J. Larner, and G. T. Cori. 1952. "Structure of Glycogens and Amylopectins. 1. Enzymatic Determination of Chain Length." *J. Biol. Chem.* 199: 631–40.

Larner, J. Personal Communication.

Larner, J. 1975. "Four Questions Times Two: A Dialogue on the Mechanism of Action of Insulin Dedicated to Earl W. Sutherland." *Metabolism* 24: 249–56.

Larner, J. 1992. "Gerty Theresa Cori: August 8, 1896–October 24, 1957." *Biogr. Mem. Natl. Acad. Sci. U.S.A.* 111–35.

Larner, J. 2003. "How I Became a Biochemical Pharmacologist." *I.U.B.M.B. Life* 55: 111–13.

Larner, J., D. L. Brautigan, and M. O. Thorner. "D-Chiro-Inositol Glycans in Insulin Signaling and Insulin Resistance." 2010. *Molecular Medicine* 16: 543–52.

Larner, J., B. Illingworth, G. T. Cori, and C. F. Cori. 1952. "Structure of Glycogens and Amylopectins. II .Analysis by Stepwise Enzymatic Degradation." *J. Biol. Chem.* 199: 641–51.

Larner, J., B. J. Jandorf, and W. H. Summerson. 1949. "On the Requirement for Diphosphopyridine Nucleotide in the Aerobic Metabolism of Pyruvate by Brain Tissue." *J. Biol. Chem.* 178: 373–82.

Larner, J., and F. Sanger. 1965. "The Amino Acid Sequence of the Phosphorylation Site on Muscle Uridine Diphophoglucose Alpha-1,4-Glucan Alpha-4-Glucosyl Transferase." *J. Mol. Biol.* 11: 491–500.

Larner, J., C. Villar-Palasi, and D. J. Richman. 1960. "Insulin-Stimulated Glycogen Formation in Rat Diaphragm. Levels of Tissue Intermediates in Short-Term Experiments." *Arch. Biochem. Biophys.* 86: 56–60.

Murad, F. 1998. *Nobelprize.org. Autobiography*, available at http://www.nobelprize.org/nobel_ prizes/medicine/laureates/1998/murad-autobio.html.

Murad, F. 2004. "Discovery of Some of the Biological Effects of Nitric Oxide and Its Role in Cell Signaling." *Bioscience Reports* 24: 453–74.

Murad, F. 2006. "Shattuck Lecture. Nitric Oxide and Cyclic GMP in Cell Signaling and Drug Development." *New Eng. J. Med.* 355: 2003–11.

Rosell-Perez, M., and J. Larner. 1964. "Studiesof UDPG-Alpha-Glucan Transglucosylase. V. Two Forms of the Enzyme in Dog Skeletal Muscle and Their Interconversion." *Biochemistry* 3: 81–88.

Rosell-Perez, M., C. Villar-Palasi, and J. Larner. 1962. "Studies on UDPG-Glycogen Transglucosylase. I. Preparation and Differentiation of Two Activities of UDPG-Glycogen Transglucosylase." *Biochemistry* 1: 763–68.

Villar-Palasi, C., and J. Larner. 1960. "Insulin-Mediated Effect on the Activity of UDPG-Glycogen Transglucosylase." *Biochim. Biophys. Acta* 39: 171–73.

Villar-Palasi, C., and J. Larner. 1961. "Insulin Treatment and Increased UDPG-Glycogen Transglucosylase in Muscle." *Arch. Biochem. Biophys.* 94: 436–42.

Wall, M. A., D. E. Coleman, J. A. Iñiguez-Lluhi, B. A. Posner, A. G. Gilman, and S. R. Sprang. 1995. "The Structure of the G-Protein Heterotrimer Gi Alpha 1 Beta 1 Gamma 2" *Cell* 83: 1047–58.

20

Contributions of Barbara and David Brown

Early History and Research in the Cori Laboratory

Both David and Barbara Brown made important contributions to the Cori laboratory. David worked with Carl and later assumed some administrative duties, whereas Barbara worked with Gerty. David was born in Ely, Nevada, on June 17, 1921, where his father was involved in the ore processing business. He skipped first grade and graduated from high school at age sixteen. His family moved to California, and he took advanced courses at Pasadena Junior College prior to being admitted to CalTech as a junior. As an undergraduate, he worked with the renowned chemist, Linus Pauling, whom he described as talking a lot and engaging in many thought experiments. Brown obtained a BS and PhD from CalTech, and his PhD research was supervised by Carl Niemann and dealt with the isolation of blood group-specific proteins. He won a Merck Fellowship, which allowed him to choose any laboratory for postdoctoral work. He interviewed at several places before settling on the Cori laboratory. There, he worked initially on the phosphorylation of hexosamine by hexokinase and on the action of phosphoglucomutase on glucosamine-6-phosphate. He collaborated extensively with Barbara Illingworth in studies of polysaccharide synthesis and the enzymatic defects in various glycogen storage diseases. He then joined Charles Park, Marvin Cornblath, and William Daughaday in studies of glucose uptake by the rat diaphragm.

He supervised the work of two brilliant graduate students. One was Luis Glaser, who studied the purification and characterization of glucose-6-phosphate dehydrogenase and the enzymes that process glucosamine. Another of Glaser's projects involved the synthesis of cellulose and chitin. More details on Glaser's career are provided in chapter 24. Another outstanding graduate student mentored by David Brown was Rosalind Kornfeld née Hauk, who was born in Dallas in 1935 and grew up in Chevy Chase, Maryland. She obtained a BS in chemistry in 1957 from George Washington University and a PhD in biochemistry from Washington University.

Her thesis work with David Brown involved the purification and characterization of glycogen synthase. She stayed on for postdoctoral work and then moved to NIH as a fellow before returning to St. Louis in 1965 to work with her husband, Stuart Kornfeld. During their lifelong collaboration, they made enormous contributions to the knowledge of nucleotide sugar biosynthesis and glycan ligands for lectins. Her personal challenge was to elucidate the structure, biosynthesis, and biological roles of glycoprotein glycans in health and disease, and in this she made great advances. She retired from Washington University in 2001 due to illness and died in 2007. Her obituary was written by Varki Ajit. Stuart Kornfeld's equally spectacular career is described in chapter 24 on Luis Glaser.

Barbara née Illingworth Brown was born on May 12, 1924, in Hartford, Connecticut, where her father was employed as an insurance adjustor for the Aetna Fire Co. (now just called Aetna). When her father was transferred to Pennsylvania, she went to high school in a suburb of Wilkes Barre. She attended Smith College and graduated BA with honors in 1946. She then went to Yale for a PhD under the mentorship of the endocrinologist Jane Russell, whose specialty was growth hormone. Russell, who could never be confused with the voluptuous film star, had worked earlier with Carl Cori studying carbohydrate metabolism in hypophysectomized rats and the relationship of the anterior and posterior lobes of the pituitary to insulin sensitivity in the rat. Upon completion of her PhD in 1950, Illingworth had hoped to go to Cambridge University to study under Frank Young, who was also a specialist in growth hormone, but her stipend was unexpectedly withdrawn. She applied to Gerty Cori, who by chance had a small stipend available. Illingworth began to work closely with Gerty on the structure of glycogen and amylopectins, together with Joseph Larner. She also worked with David Brown and Carl Cori on the role of pyridoxal-5-phosphate in phosphorylase activity, the synthesis of glycogen by phosphorylase, and on identifying the enzymes deficient in various glycogen storage diseases. Much of her work with Gerty was on the enzymes involved in the characterization of glycogen branching and debranching enzymes and was done together with David. When Gerty tragically died in 1957, Illingworth carried on with her studies of the enzymatic basis for these diseases. Her work on glycogen structure and glycogen storage diseases s described in chapter 18.

She relates the various efforts to ameliorate Gerty's anemia, including splenectomy, transfusions, and various drugs—all to no avail. She describes how Gerty tried to work to the end, but often had to rest in her office. As noted by other contributors to this book, the Coris could be very intimidating to seminar speakers because of Carl's exceptional command of the literature and Gerty's ability to silently convey her impression of a talk by a look whose meaning was unmistakable. The departmental seminars were preceded by lunch with sandwiches and included visitors and some members of other departments. The seminar room was the only one air-conditioned in the department despite the oppressive summers in St. Louis. The department had only two telephones—one was for Carl and Gerty, and the other had to be shared by the others. The only cold room was a tiny one that occupied a corner of Gerty's laboratory; they bought it with funds from the Nobel Prize. In retrospect, it is amazing how much they accomplished under these conditions and how they made the most of their reagents.

Barbara and David married in 1951 and continued their collaboration with studies leading to the discovery of a new pathway for the enzymatic debranching of glycogen. This involved the combined action of oligo-1,4 → 1,4-glucantransferase and amylo-1,6-glucosidase (the transferase was earlier found by them to be a contaminant of the glucosidase). They also characterized liver lysosomal α-glucosidase. They then turned their attention to glycogen storage diseases, where many of these enzymes were absent. They demonstrated the simultaneous absence of lysosomal α-1,4-glucosidase and α-1,6-glucosidase in the type II form of the disease and the absence of branching enzyme (amylo-1,4 → 1,6-transglucosidase) in the type IV form. More studies showed the absence of debranching enzyme in the type III form of the disease. Phosphorylase was found to be deficient in muscle in type V glycogen storage disease. As described above, Joseph Larner also participated in Gerty's studies of glycogen storage diseases. In later work, Barbara collaborated with the renowned liver transplant surgeon, Thomas Starzl at the University of Pittsburgh, to see if transplantation of patients with glycogen storage diseases could ameliorate the condition.

David was an important collaborator with Barbara, but he was also involved in other important duties at Washington University. In 1950, he assumed administrative responsibilities for the department, thus relieving Carl Cori of duties he did not enjoy. David's responsibilities included directing the graduate program, and later extended to the whole medical school, where one of his major duties was to supervise the design, construction, and financial arrangements of two major additions to the medical center. One was Queeny Tower, which provided floors for hospital rooms, office space, hotel-type accommodation for families of patients, and a better quality restaurant with funds from Edgar Queeny, whose family founded the Monsanto Company. Another building whose construction David was involved with was the McDonnell Science Building, a gift from James S. McDonnell, whose company was engaged in aircraft construction (McDonnell-Douglas) in St. Louis. David continued to be active in research, principally in association with Barbara, with a focus on enzymes involved in glycogen metabolism and various glycogen storage diseases, as noted above. He partly retired in 1986, but continued to teach first-year medical students until 1989 when Barbara also retired. He had a longstanding interest in plants and, as a teenager, had collected all the plant specimens in his county in Nevada—his herbarium remains in good condition to this day. His interest in plants continued in St. Louis, where he spent his time growing orchids, of which he became an expert. He later became president of the National Orchid Society. David Brown died on June 13, 2011, after a prolonged illness.

References

Ajit, V. 2007. "Rosalind Hauk Kornfeld (1935–2007)." *Glycobiology* 17: 1148–49.
Brown, B. I. Personal Communication.
Brown, D. H. Personal Communication.
Brown, B. I., D. H. Brown, and P. L. Jeffrey. 1970. "Simultaneous Absence of Alpha-1,4-Glucosidase and Alpha-1,6-Glucosidase Activities in Tissues of Children with Type III Glycogen Storage Disease." *Biochemistry* 9: 1423–28.
Brown, D. H., B. Illingworth, and C. F. Cori. 1963. "Combined Action of Oligo-1,4 → 1,4-Glucantransferase and Amylo-1,6-Glucosidase in Debranching Glycogen." *Nature* 197: 980–82.
Chen, Y. T., J. K. He, J. H. Ding, and B. I. Brown. 1987. "Glycogen Debranching Enzyme: Purification, Antibody Characterization, and Immunoblot Analyses of Type III Glycogen Storage Disease." *Am. J. Hum. Genet.* 41: 1002–15.
Gibson, W. B., B. Illingsworth, D. H. Brown. 1971. "Studies of Glycogen Branching Enzyme. Preparation and Properties of 1,4-Glucan → 1,4Glucan 6-Glycosyl Transferase and Its Action on the Characteristic Polysaccharide of the Liver of Children with Type IV Glycogen Storage Disease." *Biochemistry* 10: 4253–62.
Illingworth, B., H. S. Jansz, D. H. Brown, and C. F. Cori. 1958. "Observations on the Function of Pyridoxal-5-Phosphate in Phosphorylase." *Proc. Natl. Acad. Sci. U.S.A.* 44: 1180–91.
Illingworth, B., J. Larner, and G. T. Cori. 1952. "Structure of Glycogens and Amylopectins. I. Enzymatic Determination of Chain Length." *J. Biol. Chem.* 188: 631–40.
Jeffrey, P. L., D. H, Brown, and B. I. Brown. 1970. "Studies of Lysosomal Alpha-Glucosidase. I. Purification and Properties of the Rat Liver Enzyme." *Biochemistry* 9: 1403–15.
Larner, J., B. Illingworth, G. T. Cori, and C. F. Cori. 1952. "Structure of Glycogens and Amylopectins. II. Analysis by Stepwise Enzymatic Degradation." *J. Biol. Chem.* 199: 641–51.

21

William Daughaday

All About Growth

Training in Endocrinology

William (Bill) Daughaday was a member of the group that studied the effects of hormones on muscle carbohydrate metabolism under the tutelage of Carl Cori. Daughaday later became one of America's leading endocrinologists and a world authority on growth hormone (Figure 20). He was born in Chicago on February 12, 1918, and attended Harvard College as a National Scholar and then Harvard Medical School, from which he graduated in 1943. His interest in endocrinology went back to his high school days. The father of one of his closest friends was Paul Starr, who was head of endocrinology at Northwestern Medical School. Starr would have liked to have pursued a full-time academic career, but because of the Great Depression, he was forced to support himself by private practice. However, he did maintain a laboratory, which Daughaday visited several times and worked in after his first year in medical school. After World War II, Starr became head of the Department of Medicine at the University of Southern California and served as president of the Endocrine Society.

When Daughaday entered Harvard College in 1936, he was soon drawn to the Biology Department since it was headed by a very productive endocrinologist whose laboratory attracted many future leaders in the field, including Roy Greep, Edward (Ted) Astwood, and Alexander Albert. Daughaday also spent two summers at the Jackson Laboratory in Bar Harbor, Maine. At that time, it was one of the few centers for mammalian genetic research, and it provided a wonderful course on genetics for the summer students. His research there was based on the observation that neonatal castration resulted in the late development of adrenal adenomas with virilizing effects in certain strains of mice. Daughaday took this project back to Harvard for his honors thesis.

He entered Harvard Medical School in 1940, a time he describes as the golden age of clinical endocrinology. His teachers at the Massachusetts General Hospital included James Means, an expert on the thyroid and the renowned Fuller Albright whose area was the parathyroid and bone disease, and Oliver Cope, a parathyroid surgeon. At the Peter Bent Brigham Hospital, the staff included Edward Astwood, with many interests in the pituitary and thyroid, and George Thorn, whose focus was on adrenal diseases. One of Daughaday's important interactions was with Robert H. (Bob) Williams, a young assistant professor at the Boston City Hospital who exhibited great enthusiasm for endocrinology. After graduation from medical school in 1943, Daughaday had a short internship at the City Hospital, followed by twenty months of service as a medical officer in the U.S. Army. In 1946–1947, he was a research fellow at the Thorndike Laboratory, where he worked in the Williams laboratory to develop an improved assay for corticosteroids in urine.

He was then recruited to the Barnes Hospital and Washington University in St. Louis by his medical school friend, Robert Glaser, who was chief resident on the very selective Ward Medical Service. Glaser persuaded Barry Wood, chairman of the Department of Medicine, to give Daughaday an appointment as an assistant resident on this service. Daughaday's job was to assist part-time faculty in the radioactive iodine treatment program for patients with hyperthyroidism (Graves disease). In the following year, he was made a fellow and had essentially complete responsibility for the radioiodine program, doing all the urinary excretion measurements in the laboratory of Martin Kamen. Kamen was a physicist who co-discovered ^{14}C while working under Ernest Lawrence in the Radiation Laboratory at Berkeley. Kamen returned to Berkeley after working on the Manhattan Project, but was fired after being accused of leaking nuclear weapon secrets to the Russians and was summoned to appear before the notorious House Committee on Un-American Activities. Because of this, he was unable to obtain an academic position until he was hired by Arthur Compton to run the cyclotron at Washington University. Compton was the chancellor of Washington University and had won the Nobel Prize in 1923 for the "Compton Effect."

At the suggestion of Barry Wood, in 1949, Daughaday sought training in biochemistry in the Cori laboratory. He was supported there as an NIH fellow and worked on the hormonal regulation of glucose metabolism of the rat diaphragm. This project was carried out with many young investigators and is described in detail in chapter 15. Daughaday was particularly interested in the effect of growth hormone and had to start by purifying the hormone from bovine pituitaries. His research was initially directed to seeing if growth hormone had an effect on hexokinase—this was during the era when the Coris believed that insulin acted by stimulating this enzyme. However, he and others in the group found no effect of growth hormone. Mike Krahl, a member of the team, thought they would get positive results if they used rats from the Anheuser Busch Company, which was more notable for its production of beer. These animals were notoriously fierce and leaped to attack anyone who approached their cages. Daughaday was assigned to inject them, at risk of life and limb, but, as experimental animals, they proved to be no better than their benign companions!

Barry Wood's plan was to bring Daughaday and Charles Park back to head the Metabolism Division in his department. Daughaday returned to the Division in 1950 as acting head but continued his research with Cori. When Park took the chair of physiology at Vanderbilt, Daughaday was made director of the Division at age thirty-three. The division consisted of a single laboratory with a small office across from an eight-bed metabolic ward in Barnes Hospital. Unfortunately, funds to support this research disappeared during the Great Depression, and it became merely a section of the medical service. Daughaday's research continued to involve radioactive iodine treatment of hyperthyroidism and measurement of adrenal steroid metabolites in urine.

With modest support from the Nutrition Foundation, he joined with Joseph Larner to explore why patients with uncontrolled diabetes excreted massive amounts of *myo*inositol in their urine. They found, using rats and humans, that there was an active renal tubular transport mechanism for the inositol that was competed against by glucose. Hence, the high glucose of the diabetics inhibited the renal reabsorption of the inositol. Later, they showed that *myo*inositol could be synthesized from glucose. When the Wohl Hospital was completed in 1953, the Metabolism Division moved into new laboratories on the eighth floor. Another research focus led to the discovery that certain plasma proteins (globulins) bound adrenal cortical hormones (corticosteroids) and were responsible for carrying them through the bloodstream. There were also clinical studies of the treatment of diabetes and diabetic ketoacidosis.

Research on Growth Hormone and Somatomedin

Daughaday's research increasingly turned to studies of growth hormone secretion and its levels and actions in acromegaly, dwarfism, diabetes, and obesity. At an early stage of the research, he found that a convenient assay of the hormone was to measure its effect on the uptake of ^{35}S-sulfate into cartilage from hypophysectomized rats. Although this was good for assaying plasma, surprisingly, addition of pure growth hormone had no effect. This led to the novel hypothesis, developed with a clinical fellow William Salmon, that the stimulation of cartilage growth was due to the fact that plasma contained a "sulfation factor" whose presence was dependent on growth hormone. This was a real advance in understanding the action of growth hormone, but was not universally accepted. However, it was supported by measurements of the factor in plasma from patients with high or low levels of growth hormone, such as acromegaly and hypopituitarism. Sulfation factor was renamed somatomedin, and two species (somatomedin A and C) were identified by Judson Van Wyk working at the University of North Carolina. In an interesting convergence of research, a Swiss investigator, Rudolf Froesch, isolated an insulin-like substance from serum. This was not blocked by insulin antibodies and was termed nonsuppressible insulin-like activity (NSILA). On purification, NSILA fractions were observed to resemble somatomedin peptides isolated by Van Wyk. Furthermore, NSILA added to rat cartilage showed somatomedin activity, and somatomedin was shown to exhibit NSILA activity on fat pads. Two workers in Zurich then purified and determined the amino acid sequence of the two components of NSILA. They were renamed insulin-like growth factors I and II (IGF-I and IGF-II), and their structural similarity to insulin was recognized. Many tissues are able to synthesize and secrete IGF I in a growth hormone-dependent manner, but Daughaday showed that the principal source is the liver, Perfused livers from hypophysectomized rats produced less IGFI, and treatment of the rats with growth hormone restored the production to normal.

Clinical Studies of Humans with Growth Disorders

As noted above, Daughaday became director of the Metabolism Division in 1951, and in 1975 he was appointed director of the Diabetes Endocrinology Research Center, which was later renamed the Diabetes Research and Training Center. He was named to an endowed professorship in 1983—the Karl Professorship in Endocrinology and Medicine. During this period, he continued clinical and basic studies of growth hormone and extended these to look at the regulation and actions of two related hormones, prolactin and placental lactogen. The clinical studies involved patients with acromegaly, gigantism, hypopituitarism, and dwarfism and required the development of radioimmunoassays for growth hormone and prolactin, and assays for IGF-I and IGF-II. In general, the plasma levels of growth hormone and IGF-I correlated with the changes in stature, but there were some striking exceptions. In the condition of Laron's dwarfism, the growth hormone level was normal and the defect was traced to a deficiency of growth hormone receptors. These patients had low levels of IGF-I, but responded to injected IGF-I, as expected. In other cases of dwarfism, the response to IGF-I was less than normal, and it was postulated that there was a defect in its receptors.

These clinical studies were accompanied by investigations into the disposition of injected growth hormone and placental lactogen in humans and also studies of the regulation of the secretion of growth hormone and prolactin conducted in humans and rats. Further work revealed the presence in plasma of a growth hormone binding protein, and they deduced that this was derived from tissue growth hormone receptors. African pygmies were found to have low levels of the binding protein, and it was surmised that this probably reflected low levels of the receptor. Besides studying patients with disorders of growth hormone levels and actions, Daughaday investigated the hormonal changes in a variety of conditions, including pregnancy, galactorrhea, diabetes, alcoholism, cirrhosis, diabetes, prematurity, and obesity.

Despite the strong evidence supporting the idea that somatomedin (IGF-I) was the mediator of growth hormone action on bone cartilage, it was not universally accepted Consequently, Daughaday had to provide additional support and write more review articles. He pointed to numerous studies in humans, for example, certain familial dwarfs or patients of low stature where growth was obviously stunted. In these subjects, growth hormone levels were normal or high, but somatomedin levels were low, or tissue responses to it were decreased. He also expanded his research on IGFI and IGF-II. For example, he developed radioimmunoassays for these and characterized their separate receptors. He identified and biochemically characterized serum binding proteins for both of the factors. Umbilical cord blood and blood from neonates were found to have lower levels of IGF-I than blood from adults. The blood level of IGF-I was found to rise during childhood and to peak during puberty, that is, during linear growth. Other studies showed that some patients with short stature had very high levels of IGF-I binding protein, and it was deduced that these interfered with the binding of IGF-I to its receptors.

Daughaday continued a tradition of the Cori laboratory, namely brown bag lunches with a free exchange of scientific ideas and a lively discussion of world and cultural affairs. Like the Cori laboratory, space at Daughaday's division was limited, and this aided communication. Yelling between the various laboratories was considered an acceptable practice. In 1972, Daughaday served as interim chair of the department and was succeeded by three endocrinologists as chairs of internal medicine—namely, David Kipnis, whose career is described in a separate chapter, Gustav Schonfeld, and Kenneth Polonsky. Philip Cryer and Clay Semenkovich led the Metabolism Division following Daughaday. All of these investigators made outstanding contributions to the understanding of diabetes and to the training of fellows in endocrinology.

Daughaday received many awards for his work. He was president of the Endocrine Society in 1971–1972, and in 1975, he received the Koch Medal, its highest honor. He was elected to the National Academy of Sciences in 1986 and to the American Academy of Arts and Sciences in 1989. He served on the editorial boards of many journals and was an associate editor of the *Journal of Clinical Investigation*. He also served on many NIH review groups and advisory committees and was on the board of Scientific Counselors of the National Institute of Arthritis and Metabolic Diseases.

References

Beck, P., and W. H. Daughaday. 1967. "Human Placental Lactogen: Studies of Its Acute Metabolic Effects and Disposition in Normal Man." *J. Clin. Invest.* 46: 103–10.

Beck, P., M. L. Parker, and W. H. Daughaday. 1965. "Radioimmunologic Measurement of Human Placental Lactogen in Plasma by a Double Antibody Method during Normal and Diabetic Pregnancies." *J. Clin. Endocrinol. Metab.* 25: 1457–62.

Daughaday, W. H. 1956. "Binding of Corticosteroids by Plasma Proteins. I. Dialysis Equilibrium and Renal Clearance Studies." *J. Clin. Invest.* 35: 1428–33.

Daughaday, W. H., *Biographical Information*, courtesy Philip Skroska, Archivist, Becker Medical Library. Washington University School of Medicine.

Daughaday, W. H., and J. Larner. 1954. "The Renal Excretion of Inositol in Normal and Diabetic Human Beings." *J. Clin. Invest.* 33: 326–32.

Daughaday, W. H., K. A. Parker, S. Borowsky, B. Trivedi, and M. Kapadia. 1982. "Measurement of Somatomedin-Related Peptides in Fetal, Neonatal, and Maternal Rat Serum by Insulin-Like Growth Factor (IGF) I Radioreceptor Assay (RRA), and Multiplication-Stimulating Activity RRA after Acid-Ethanol Extraction." *Endocrinology* 110: 575–81.

Daughaday, W. H., W. D. Salmon Jr., and F. Alexander. 1959. "Sulfation Factor Activity of Sera from Patients with Pituitary Disorders." *J. Clin. Endocrinol. Metab.* 19: 743–58.

Daughaday, W. H., B. Trvedi, and M. Kapadia. 1981. "Measurement of Insulin-Like Growth Factor II by a Specific Radioreceptor Assay in Serum of Normal Individuals, Patients with Abnormal Growth Hormone Secretion, and Patients with Tumor-Associated Hypoglycemia." *J. Clin. Endocrinol. Metab.* 53: 289–93.

Elders, M. J., J. T. Garland, W. H. Daughaday, D. A. Fisher, J. E. Whitney, and E. R. Hughes. 1973. "Laron's Dwarfism: Studies on the Nature of the Defect." *J. Pediatr.* 83: 253–63.

Jacob, A., C. Hauri, and E. R. Froesch. 1968. "Nonsuppressible Insulin-Like Activity in Human Serum. 3. Differentiation of Two Distint Molecules with Nonsuppressible Insulin-Like Activity." *J. Clin. Invest.* 47: 2678–2688.

Jacobs, L. S., I. K. Mariz, and W. H. Daughaday. 1972. "A Mixed Heterologous Radoimmunoassay for Human Prolactin." *J. Clin. Endocrinol. Metab.* 34: 484–90.

Kamen, M. 1985. *Radiant Science, Dark Politics*. Berkeley: University of California Press.

Merimee, T. J., G. Baumann, and W. H. Daughaday. 1990. "Growth Hormone-Binding Protein: Studies in Pygmies and Normal Statured Subjects." *J. Clin. Endocrinol. Metab.* 71: 118.

Parker, M. L., R. D. Utiger, and W. H. Daughaday. 1962. "Studies of Human Growth Hormone. II. The Physiological Disposition and Metabolic Fate of Human Growth Hormone in Man." *J. Clin. Invest.* 41: 262–68.

Phillips, L. S., A. C. Herington, I. E. Karl, and W. H. Daughaday. 1976. "Comparison of Somatomedin Activity in Perfusates of Normal and Hypophysectomized Rat Livers with or without Added Growth Hormone." *Endocrinology* 98: 606–14.

Rinderknecht, E., and R. E. Humbel. 1978. "The Amino Acid Sequence of Human Insulin-Like Growth Factor 1 and Its Structural Homology to Proinsulin." *J. Biol. Chem.* 253: 2769–76.

Salmon, W. D. Jr., and W. H. Daughaday. 1957. "A Hormonally Controlled Serum Factor which Stimulates Sulfate Incorporation by Cartilage in vitro." *J. Lab. Clin. Med.* 49: 825–36.

Scott, M. G., G. C. Cuca, J. R. Peterson, L. R. Lyle, B. D. Burleigh, and W. H. Daughaday. 1987. "Specific Immunoradiometric Assay of Insulin-Like Growth Factor I with Use of Monoclonal Antibodies." *Clin. Chem.* 33: 2019–23.

Tollefsen, S. E., E. Heath-Monnig, M. A. Cascieri, M. L. Bayne, W. H. Daughaday. 1991. "Endogenous Insulin-Like Growth Factor (IGF) Binding Proteins Cause IGF-1 Resistance in Cultured Fibroblasts from a Patient with Short Stature." *J. Clin. Invest.* 87: 1241–50.

Van Wyk, J. J., K. Hall, J. L. van den Brande, and R. P. Weaver. 1971. "Further Purification and Characterization of Sulfation Factor and Thymidine Factor from Acromegalic Plasma." *J. Endocrinol. Metab.* 32: 389–403.

22

Robert Crane

A Decade with Carl Cori

Early Research in Boston

Robert Crane spent the period 1950 to 1962 in the Cori department. He was endowed with great self-confidence and never hesitated to challenge any concept or person, including Carl Cori. His granduncle was Stephen Crane, author of the famous Civil War novel, *Red Band of Courage*. He attended St. Andrews Episcopal High School in Delaware, where he familiarized himself with H. W. Fowler's *A Dictionary of Modern English Usage*, which gave him a thorough grounding in grammar that he appreciated all his life. In 1938, he entered Washington College, a liberal arts college on the eastern shore of Maryland. His brother had chosen to go there the previous year, because during a visit to the campus, he had noticed a strikingly beautiful girl with shoulder-length blonde hair. Freshmen were required to take one laboratory science course, and Crane tossed a coin that landed for chemistry. However, he was in luck, because the chemistry teacher generated much interest in the subject. Crane also became impressed with biochemistry as the route to solving biological problems. Upon graduation, he was accepted by E. V. McCollum at Johns Hopkins, but McCollum would not give him a draft deferment. So he trained for TNT production at a plant in Pennsylvania. He worked as an analytical chemist, doing research in one of the dynamite mixing houses. These were surrounded by heavy earthworks, and the one he worked in blew up the next year!

Crane then spent a year teaching chemistry in a teachers college in Missouri and was drafted into the U.S. Navy. He entered boot camp twice and was assigned to various naval facilities on the East Coast. At one of these in Florida, he became seriously ill, and it took a long time to make the diagnosis, which was meningococcal septicemia. Luckily, there was enough penicillin available to treat this potentially fatal disease, and he was back to normal in a few days. He went back to the navy to stations on the West Coast, serving on a destroyer at Pearl Harbor, the Philippines, Borneo, and Japan. He didn't get an expected promotion and left the Navy.

Crane's wife urged him to study biochemistry under the auspices of the GI bill. So he went to Harvard to study under Eric Ball, who was chairman of the division of medical sciences and an expert on the metabolism of adipose tissue. Ball was famous as the discoverer of the antilipolytic action of insulin, which is important in regulating the breakdown of triglyceride to fatty acid. In diabetes, the increased release of fatty acids from adipose tissue is the major contributor to ketoacidosis. Ball set Crane to work making crystalline yeast hexokinase—a venture that was not successful. Nevertheless, he was sent to the Marine Biological Laboratory at Woods Hole, Massachusetts, to work with G. H. A. Clowes. There he met Mike Krahl, a professor in the Cori department, and Krahl later persuaded Carl Cori to accept Crane into his department. Back at Harvard, Ball required him to study CO_2 fixation in the retina. Ball's method of training graduate students was the "sink or swim" variety, and Crane's ideas often conflicted with Ball's.

He next went to Fritz Lipmann's laboratory at the Massachusetts General Hospital in 1949, seven months before the award of his PhD. He wasn't sure why Lipmann accepted him. While most people there were on the trail of CoA, he and Hermann Niemeyer worked on oxidative phosphorylation. In his autobiography, Crane said that Ball taught him how to do research, but Lipmann taught him how to think. Nearly every afternoon, Lipmann would visit Crane's tiny laboratory and sit down and talk. His advice for the way to do research was to "follow your nose." In 1950, it was conventional wisdom that 1,3-bisphosphoglycerate was produced by the oxidation of a complex between 3-phosphoglyceraldehyde and inorganic phosphate, a concept originated by Otto Warburg and perpetuated by Lipmann. Crane proposed a contrary view that oxidative formation of an acyl-enzyme followed by phosphorolysis was more likely.

Research in the Cori Laboratory

In September of 1950, Crane went to the Cori laboratory to test his hypothesis of how 1,3-bisphosphoglycerate was formed. There he encountered Sidney Velick (see chapter 9) and Velick's graduate student Jane Harting (see chapter 17), who were deeply involved in studying the mechanism of action of glyceraldehyde-3-phosphate dehydrogenase. Harting was working in the next laboratory and agreed to test his hypothesis, but there was a question of continuing the project due to lack of progress. He persuaded her to continue and conduct a decisive experiment, which proved the correctness of his hypothesis and gave her an elegant thesis. Crane continued to work on aspects of oxidative phosphorylation as they related to the mitochondrial membrane, but could make little sense of the results. Carl Cori then suggested that he work on hexokinase, and in this he was joined by Alberto Sols, whose career is described in chapter 23. At this time (1951), Cori still believed that insulin acted on this enzyme and not on glucose transport. This was despite the fact that Rachmiel

Levine had found evidence that the hormone acted on glucose penetration. Crane attributed this to the fact that it takes a few years for a revolutionary concept to be appreciated, tested by others, and finally accepted. During this period, Cori was opposed to research on membrane transport.

Crane's work with Sols on hexokinase involved studies of its substrate specificity, inhibition by glucose-6-phosphate and ADP, and its binding to a particulate fraction from brain. In his autobiography, he said that he preferred to do experiments to answer specific questions or to test a hypothesis, and shied away from just data collection. However, he recognized that sometimes data collection led to unexpected findings. He and Sols incorporated their findings with hexokinase into a general scheme whereby the concentration of glucose-6-phosphate could regulate glycolysis, glycogenolysis, and glycogen synthesis. Later, they recognized that the inhibitory effect of glucose-6-phosphate on hexokinase was an example of allosteric regulation.

Discovery of Glucose-Na$^+$ Cotransport

For his next research project, Crane turned to the phenomenon of membrane transport. The test system was Ehrlich ascites tumor cells, which were obtained from Paul Zamecnik's laboratory at the Massachusetts General Hospital, and they examined the penetration of sugars. The process showed reversibility, temperature dependence, and substrate specificity. Competitive inhibition between the sugars was observed, and insulin was found to have no effect. Cori and Crane discussed the results almost every day and speculated about the possible mechanisms by which the transport might work. In his autobiography, Crane notes that the value to his personal growth of the long, close relationship with Carl Cori during this period cannot be overestimated. While Helmreich, Kipnis, Park, and others focused on the metabolic consequences of changes in glucose transport in muscle, Crane explored the mechanism of active transport using the inverted intestinal sac preparation, with emphasis on the possible role of phosphorylation. The idea set forth by Lundsgaard that transport involved a phosphorylation-dephosphorylation mechanism had become questionable, but Crane, using different glucose analogs, decisively disproved the hypothesis. For some reason, Carl Cori ordered a halt to the work—an uncharacteristic act—and this was very disturbing to Crane. As a result, he moved on to study galactose uptake by kidney slices. This work showed that the glycoside phlorizin inhibited the active transport of glucose by acting on the membrane and not on an intracellular phosphorylation reaction.

Studies of active transport exploring the exchange of ^{18}O with the carbon-2 hydroxyl of glucose and carried out with Mildred Cohn and George Drysdale indicated that a covalent reaction with glucose was not the basis for active transport. The next step was to find out at which border of the intestinal cell (brush border or basolateral domain) the gradient of glucose was established. His initial efforts were unsuccessful, but a collaboration with Oliver Lowry, known for his expertise in micromethods, showed it to be the brush border.

The next issue Crane explored was the nature of the system that provided energy to the carrier. He postulated that this was a Na^+ circuit established by an ATP-energized pump and did not involve phosphorylation-dephosphorylation. At this time, Crane became aware of the Na^+-K^+ATPase or Na^+ pump hypothesis of Jens Skou and realized that a Na^+ gradient could provide the driving force for glucose transport. Skou won the Nobel Prize in 1997 for the discovery of the Na^+ pump.

The issue then became how to link the influx of Na^+ at the brush border membrane to the transport of sugar. He found that Na^+ was absolutely required for glucose entry, but not Cl^-. He developed the concept of glucose-Na^+ cotransport by a mobile carrier mechanism in which there was no chemical reaction of glucose. The requirement for ATP (to drive the Na^+ pump) was shown through the use of uncouplers of oxidative phosphorylation or replacement of O_2 with N_2. Furthermore, the level of accumulated sugar varied with the external concentration of Na^+. Crane's concept of ion-coupled membrane transport was first drawn out in the sand of a beach at Woods Hole, Massachusetts, and then more formally presented at the International Conference on Membrane Transport and Metabolism at Prague in August 1960. Because of the limits of time, Crane could present only a few critical experiments, but he knew that the gradient-coupled cotransport model was the only one to fit the criteria. In fact, he either knew or had reason to believe that his hypothesis fitted ten criteria. When he finished his talk, Peter Mitchell cried out, "You've got it." It was Crane's opinion that the modern era of the biochemistry of membrane transport was born at that meeting and that it led on to the chemiosmotic theory of oxidative phosphorylation promulgated by Peter Mitchell, winning him the Nobel Prize in 1978.

Crane's discovery of cotransport led to the development of oral rehydration, which has been responsible for saving the lives of millions of patients suffering from diarrhea, especially those in developing countries with diarrhea caused by cholera. The procedure involves oral administration of a solution of salts and sugars and is undoubtedly Crane's greatest contribution to medicine. His work also led to the antidepressant fluoxetine (Prozac), which acts by blocking the uptake of serotonin by the brain through action on Na^+-serotonin cotransporters. He left the Cori department in 1962 to become chairman of the Biochemistry Department at the Chicago Medical School. When he left St. Louis, Carl told him, "Bob, it was very difficult living with you, but immensely worthwhile." In 1966, Crane became chairman of the Department of Physiology and Biophysics at Rutgers Medical School (now known as the Robert Wood Johnson Medical School). He was elected a fellow of the American Association for the Advancement of Science and received a Distinguished Achievement Award from the American Gastroenterological Association in 1969 and a ScD from Washington College in 1982. On his retirement from the chair at Rutgers, he purchased a thirty-five-foot fiberglass sloop and lived in it for more than five years, spending the summers sailing along the East Coast as far as Cape Cod. He then settled near Memphis, Tennessee, where he had several horse farms. Crane died in Somerville, Tennessee, on October 31, 2010, at the age of ninety.

References

Cohn, M. 1956. "Some Mechanisms of Cleavage of Adenosine Triphosphate and 1,3-Diphosphoglycerate." *Biochim. Biophys. Acta* 20: 92–99.
Crane, R. K. 1960. "Intestinal Absorption of Sugars." *Physiol. Rev.* 40: 789–825.
Crane, R. K. 1983. "The Road to Ion-Coupled Membrane Processes." *Comprehensive Biochemistry* 35: 43–69.
Crane, R. K. 2009. *Living History of Physiology Project*. Edited by M. Frank, Bethesda: American Physiological Society.
Crane, R. K., and E. G. Ball. 1951. "Relationship of $C^{14}O_2$ Fixation to Carbohydrate Metabolism in Retina." *J. Biol. Chem.* 189: 269–76.
Crane, R. K., and F. Lipmann. 1953. "The Relationship of Mitochondrial Phosphate to Aerobic Phosphate Bond Generation." *J. Biol. Chem.* 201: 245–46.
Crane, R. K., and A. Sols. 1953. "The Association of Hexokinase with Particulate Fraction of Brain and Other Tissue Homogenates." *J. Biol. Chem.* 203: 273–92.
Crane, R. K., and A. Sols. 1954. "The Non-competitive Inhibition of Brain Hexokinase by Glucose-6-Phosphate and Related Compounds." *J. Biol. Chem.* 210: 597–606.
McDougal, D. B. Jr., K. D. Little, and R. K. Crane. 1960. "Studies on the Mechanism of Intestinal Absorption of Sugars. IV. Localization of Galactose Concentrations within the Intestinal Wall during Active Transport in vitro." *Biochim. Biophys. Acta* 45: 483–89.
Skou, J. C. 1957. "The Influence of Some Cations on an Adenosine Triphosphatase from Peripheral Nerves." *Biochim. Biophys. Acta* 23: 394–401.
Sols, A., and R. K. Crane. 1954. "The Inhibition of Brain Hexokinase by Adenosine Diphosphate and Sulfhydryl Reagents." *J. Biol. Chem.* 206: 925–36.
Sols, A., and R. K. Crane. 1954. "Substrate Specificity of Brain Hexokinase." *J. Biol. Chem.* 210: 581–95.

23

Alberto Sols

Spanish Enzymologist

Education in Spain

Alberto Sols Garcia spent the period 1951 to 1953 in the Cori laboratory. As indicated in the previous chapter, he worked closely with Robert Crane in studies of hexokinase. Sols was born in the city of Sax, in the province of Alicante, Spain, on February 2, 1917. His father, Pedro Sols Lluch, was a solicitor, and his mother, Maria del Amor Garcia Martinez, was the daughter of a physician. By error, neither his father nor the priest registered his birth, and, lacking a birth certificate, Sols was unable to enroll in school. Thanks to his brother who was studying law and a doctor who delivered his father, a late registration of his birth was allowed. He started in a preschool of the Carmelite Sisters, but his father was concerned that the sisters did not worry enough about academic progress, but instead taught him doctrine and prayers with the aim of educating a pious boy with good feeling and discretion.

When Sols was nine years old, the family moved to Mogente, in the province of Valencia, but his father fell ill with prostatitis and, despite a move to the city of Valencia and surgery, his father died at age sixty-one. This left the family in a precarious financial situation, and Sols was unable to continue at a private Jesuit school. They moved to Chelva to live with his mother's brothers. He was able to become a boarder at a Jesuit college in San Jose because his brother had reported the financial problems of the family, and the college provided a scholarship funded by one of the alumni with the strict proviso that the donor remain anonymous. Sols stayed at this school until 1932. He continued his studies at the Institute for Media Education Luis Vives in Valencia. However, because of continuing family financial problems, he moved to the College of the Blessed John de Ribera in Bujassot, Valencia, where he got free accommodation, tuition, and books.

In 1935, he began studies in the Faculty of Medicine at the University of Valencia. He had debated between studying chemistry and philosophy, but his interest in biology guided him to medical studies, because this would lead to knowledge of the chemical basis of life. When the Civil War broke out in Spain, he and his brother enlisted in the army. He received books in mathematics to continue his studies while he trained as a lieutenant. Between May and September of 1937, he was in the area of Navas del Marqués (Ávila), and in October 1937, he was assigned as a lieutenant to a heavy artillery unit in Medina del Campo (Valladolid) and then he went to the front at Teruel. He was a member of the Opus Dei, and this religious organization had a significant influence upon him at this point in his life. Opus Dei (Work of God) is a controversial organization in the Catholic Church that has been criticized for its secretiveness, elitism, and support of right-wing causes, including the government of Francisco Franco.

At the end of the Civil War, Sols returned to his medical studies and received his doctorate in Madrid in 1944. He decided to become a university professor and went to Barcelona, where was appointed an assistant in the medical research laboratory in the Faculty of Medicine there. In order to enhance his career, he decided to publish extensively in physiological journals and was secretary of the *Spanish Journal of Physiology*, founded in Barcelona in 1945. The journal tried to provide a venue for international dissemination of the work of Spanish scientists in the fields of normal and pathological physiology and biochemistry. In 1947, Sols was given a four-year renewable assistant professorship of physiology at the medical school. He started making contacts with foreign scientists and decided to go to the United States to complete his scientific education.

Research on Hexokinase in the Cori Laboratory

Sols chose the Cori laboratory for his studies because both Coris had won the Nobel Prize, and their research was at the forefront of biochemistry. As described in the preceding chapter, Sols worked with Robert Crane on hexokinase, the first enzyme in the glycolytic pathway. The first discovery was surprising, namely, that almost all the hexokinase in brain was not soluble but was associated with a particulate fraction. There was also significant association of hexokinase with particulate elements in other tissues. Low concentrations of glucose-6-phosphate inhibited hexokinase from all sources, and brain hexokinase was also inhibited by ADP. In addition, the enzyme was inhibited by reagents that react with free sulfhydryl groups, indicating the requirement of these groups for activity. The Michaelis constant (K_m), which is the substrate concentration for half-maximal velocity of an enzyme reaction, was determined for a variety of sugars that were substrates for brain hexokinase. The best substrates included glucose, mannose, 2-deoxyglucose, glucosamine, and N-acetylglucosamine, and it was deduced that the specificity of the enzyme involved the ring structure and the hydroxyl groups at carbons 1, 3, 4, and 6 of the glucose molecule. Some of the compounds were found to act as competitive inhibitors. The inhibitory action of glucose-6-phosphate and related compounds was also explored and found to be noncompetitive. Analysis of the structures of these inhibitors also indicated the need for specific structural requirements. All of these studies were important, because after glucose transport, phosphorylation of glucose was the next critical step in the utilization of glucose as a fuel by tissues.

Return to Spain

Sols returned to Spain in 1954 to the Department of Enzymology of the Centro de Investigaciones Biológicas in Madrid. There he continued his research on carbohydrate metabolism with studies on hexokinase in the pancreas and honey bees. Because of limitations in the support of research by Spanish scientists, Oliver Lowry's wife Norma arranged for funds to be sent to Sols from Washington University. He continued to send papers to scientific journals in the United States, Britain, and Europe, and their publication enhanced his international reputation. In 1956, he went to a New Year's Eve party at the home of the director of the Institute of Metabolism, Jose Luis Rodriguez-Candela. It was at this event that he met the director's sister, Angelines Rodriguez-Candela, who later became his wife.

In May 1960, the Instituto Marañón—which included the Departments of Metabolism and Nutrition and the Laboratory of Enzymology—was established in Madrid. Rodriguez-Candela was appointed director and Sols deputy director. Sols continued his work on hexokinase, making the important observation that, in liver, there is a second enzyme that phosphorylates glucose with different kinetic properties, and that the other enzyme (glucokinase) is involved in glycogen synthesis. Glucokinase was purified, and its properties were extensively determined. Further work showed that the synthesis of glucokinase in the liver was decreased by diabetes and fasting, and it was shown that its synthesis was controlled by insulin. Sols' research was extended to other areas of carbohydrate metabolism, for example, the transport and phosphorylation of sugars in adipose tissue.

One of Sols' significant achievements was the establishment of the Spanish Society of Biochemistry, which he accomplished with the help of Severo Ochoa, the secretary general of the Consejo Superior de Investigaciones Científicas, and several other Spanish scientists. In 1963, Sols became its first president. Sols had a strong personal friendship with Ochoa, and some of his students went to New York to work with the Nobel Laureate. Sols also began efforts to persuade Ochoa to return to Spain. These efforts would later be successful, with the establishment of the Center for Molecular Biology in Ochoa's name, which was inaugurated in 1975 and was preceded by a symposium in his honor, which was attended by ten Nobel Laureates, including Carl Cori. In chapter 14, Arnold Kornberg described the events surrounding Ochoa's seventy-fifth birthday in 1975 as a party the likes of which has not been seen in scientific circles before or since!

Sols' research interests turned to yeast carbohydrate metabolism, and his work included studies of hexose transport, glycerol metabolism, and the inhibitory effects of 2-deoxyglucose. He also studied the regulation of phosphofructokinase, fructose 1,6-bisphosphatase, and pyruvate kinase in yeast. However, he maintained his interest in the regulation of glycolysis and gluconeogenesis at the enzyme level, particularly as it applied to the liver. From this work and studies in ascites tumor cells emerged a series of insightful papers. He also contributed many useful review articles in the general area of the control of enzyme activity. Sols received many awards and honors for his contributions to science, chief among them were Academician of the Royal Society of Medicine, the National Prize for Scientific Research Ramón y Cajal (which was presented by King Juan Carlos at the Royal Palace), the Prince of Asturias Prize for Scientific Research, and awards of Doctor Honoris Causa from several Spanish universities. He served as president of the National Committee for Biochemistry (Spain) and was a founding member of the Federation of European Biochemical Societies (FEBS) and a member of the Board of the International Union of Biochemistry (IUB). He served on the executive committee of the Consejo Superior de Investigaciones Científicas and as a board member of the International Cell Research Organization. Sols died suddenly on August 9, 1989. In memoriam, he was awarded High Distinction of the Generalitat Valenciana, the Silver Medal of the National Research Council, and a medal from the Luso-Spanish Congress of Biochemistry.

References

Crane, R. K., and A. Sols. 1953. "The Association of Hexokinase with Particulate Fractions of Brain and Other Tissue Homogenates." *J. Biol. Chem.* 203: 23–292.
Crane, R. K., and A. Sols. 1954. "The Non-competitive Inhibition of Brain Hexokinase by Glucose-6-Phosphate and Related Compounds." *J. Biol. Chem.* 210: 597–606.
De La Fuente, G., and A. Sols. 1962. "Transport of Sugars in Yeast. II. Mechanisms of Utilization of Disaccharides and Related Glycosides." *Biochim. Biophys. Acta* 56: 49–62.
Gancedo, J. M., C. Gancedo, and A. Sols. 1967. "Regulation of the Concentration or Activity of Pyruvate Kinase in Yeasts and Its Relationship to Gluconeogenesis." *Biochem. J.* 102: 23C–25C.
Gancedo, C., J. M. Gancedo, and A. Sols. 1967. "Metabolite Repression of Fructose 1,6-Diphosphatase in Yeast." *Biochem. Biophys. Res. Commun.* 26: 528–31.
Gancedo, C., G. M. Gancedo, and A. Sols. 1968. "Glycerol Metabolism in Yeasts. Pathways of Utilization and Production." *Eur. J. Biochem.* 5: 165–72.
Heredia, C. F., and A. Sols. 1964. "Metabolic Studies with 2-Deoxyhexoses. II. Resistance to 2-Deoxyglucose in a Yeast Mutant." *Biochim. Biophys. Acta* 86: 224–28.
Hernández, A., and A. Sols. 1963. "Transport and Phosphorylation of Sugars in Adipose Tissue." *Biochem. J.* 86: 166–72.
Salas, J., M. Salas, E. Viñuela, and A. Sols. 1965. "Gluokinase of Rabbit Liver." *J. Biol. Chem.* 240: 1014–18.
Salas, M.,E. Viñuela, and A. Sols. 1963. "Insulin-Dependent Synthesis of Liver Glucokinase in the Rat." *J. Biol. Chem.* 238: 3535–38.
Santesmases, M. J. 1998. *Biography of Albert Sols*. Alicante, Spain: City of Sax.
Sillero, A., M. A. Sillero, and A. Sols. 1969. "Regulation of the Level of Key Enzymes of Glycolysis and Gluconeogenesis in Liver." *Eur. J. Biochem.* 10: 351–54.

Sols, A., and R. K. Crane. 1954. "The Inhibition of Brain Hexokinase by Adenosine Triphosphate and Sulfhydryl Reagents." *J. Biol. Chem.* 206: 925–36.

Sols, A., and R. K. Crane. 1954. "Substrate Specificity of Brain Hexokinase." *J. Biol. Chem.* 210: 581–95.

Salas, M., E. Viñuela, M. Salas, and A. Sols. 1965. "Citrate Inhibition of Phosphofructokinase and the Pasteur Effect." *Biochem. Biophys. Res. Commun.* 19: 371–76.

Sols, A. 1969. "Regulation of Glycolysis and Gluconeogenesis at the Enzyme Level." *Biochem. J.* 112: 2P.

24

Luis Glaser

The Complexity of Carbohydrates

Escape from Nazism and Work in the Cori Laboratory

Luis Glaser was born in Vienna on March 30, 1932. He was six years old when his family decided to leave because of the Anschluss (occupation of Austria by German troops) in 1938 and because of the prevalent pro-Nazi sentiments. They escaped to Belgium and spent almost a year there. The Belgians were extremely hospitable and allowed his father to practice some medicine and earn a little money. They left on a Dutch freighter for Mexico with tourist visas of questionable legality. On arrival at Veracruz, the authorities would not let them leave the ship, which then went on to Tampico; they would have been sent back, but a Jewish family took care of them until they could go to Mexico City several months later. Glaser went to elementary and high school there and then to the University of Toronto, where he graduated with honors in 1953. He had been fascinated by the biochemistry course and told his father he was no longer interested in medicine and wanted to go into biochemistry instead. His father was not too happy about this, but eventually became resigned to it.

In looking around for a suitable place to work on his PhD, Glaser decided to join the Cori department and was assigned to David Brown as mentor. His initial work was on the purification and characterization of glucose-6-phosphate dehydrogenase. This enzyme, which was discovered by Otto Warburg and named Zwischenferment, had been partially purified by Arthur Kornberg. Other projects were the enzymatic synthesis of hyaluronic acid chains and the synthesis of chitin, which are both glycosaminoglycans (polysaccharide chains of repeating disaccharide units containing an aminosugar). They are biologically very important, because hyaluronic acid is found in all the tissues and fluids of animals, and chitin is the main component of the exoskeleton of arthropods and is one of the most abundant biopolymers on Earth. As described in the chapter devoted to

Luis Leloir, uridine diphosphate glucose (UDPG) is involved in the synthesis of glycogen and cellulose, but there are similar compounds involved in the synthesis of other polysaccharides in which the nucleoside and sugar are different, including, for example, thymidine diphosphate glucose (TDP glucose), TDP rhamnose, and TDP aminosugars (Figure 21).

Studies with Stuart Kornfeld

Together with Stuart Kornfeld, Glaser studied the synthesis of TDP and also TDP glucose, TDP rhamnose, TDP glucosamine, and TDP N-acetylglucosamine. Kornfeld was a native of St. Louis, born on October 4, 1936. He excelled in both sports and academics but was encouraged by his parents to become a physician. He went to Dartmouth College and then obtained his MD from Washington University in 1962. As described in chapter 20 on the Browns, Kornfeld met Rosalind Hauk, who was working on her PhD with David Brown, and they married in 1959 and worked together as a team for most of their careers. After doing postdoctoral fellowships with Victor Ginsburg, an expert on glycoproteins at NIH, they returned to Washington University. Back in St. Louis, Kornfeld worked with the legendary clinician and renowned teacher, Carl Moore, and rapidly rose to the level of full professor of medicine. He continued doing research in the field of glycobiology, making major contributions to the understanding of the molecular basis of protein trafficking in mammalian cells, with focus on the targeting of lysosomal enzymes from their site of synthesis in the endoplasmic reticulum to their destination in the lysosomes. This is a multistep process involving recognition signals that must be deciphered by cell components that mediate the sorting, packaging, and transport of the enzymes. A key step is the selective phosphorylation of mannose residues on the lysosomal enzymes, because these residues allow binding to mannose-6-phosphate receptors in the Golgi and transport to lysosomes via clathrin-coated vesicles. This involved working out the entire pathway for N-linked glycan chain processing and also the entire mannose-6-phosphate pathway. Both pathways are extraordinarily complex, and Kornfeld has recently focused on the enzyme that generates the mannose-6-phosphate recognition marker on lysosomal enzymes and on the coat proteins, termed GGAs, that function as connectors to transfer the mannose-6-phosphate receptors to the clathrin-coated vesicles.

Kornfeld's career has been particularly distinguished. He has received many local and national awards, including the Passano award, the Karl Meyer Award from the Society for Glycobiology, the Kober Medal of the Association of American Physicians, and the Harden Medal from the Biochemical Society. In 1982, he was elected to the National Academy of Sciences, and this was followed by election to the Institute of Medicine and the American Academy of Arts and Sciences. He is a member of many prestigious editorial boards and has served as associate editor and then editor of the *Journal of Clinical Investigation*. In addition to his national contributions, Kornfeld provided significant service at Washington University School of Medicine as director of the Divisions of Oncology and Hematology, director of the Medical Scientist Training Program, and co-director of the Physician Scientist Training Program.

Research with Eric Olson

One of the scientists who did research with Glaser was Eric Olson, who subsequently had an impressive career. Olson grew up in North Carolina and attended Wake Forest University, where he obtained a BA in 1977 and a PhD in biochemistry in 1981. As a postdoctoral student, he wanted to apply biochemistry to questions in developmental biology. He decided to work for Glaser because he was doing exciting work in this area and Washington University had a premiere reputation. Olson's research initially was concerned with creatine phosphokinase, and he purified the brain isozyme, accomplished its cell-free translation, and studied its regulation during differentiation in the BC_3H1 muscle cell line. His next project involved studies of the regulation of the synthesis of nicotinic acetylcholine receptors, which are extraordinarily important since they convey nervous impulses in neuromuscular junctions and between neurons in components of the nervous system. In his final work with Glaser, Olson studied the control of differentiation of BC_3H1 cells by fibroblast growth factor and the fatty acylation of cellular proteins, a theme later developed by Jeffrey Gordon.

Olson states that his time in the Glaser laboratory was transformative and that Glaser provided complete freedom for his postdoctoral students to pursue their interests. Initially, this presented a challenge, but ultimately, the environment of the laboratory and Glaser's guidance enabled Olson to mature as a scientist and to develop his own independent research program. He states that he became immersed in the intensely stimulating atmosphere at Washington University and was introduced to a level of scientific intensity that he had not previously experienced. He cites Glaser as one of his role models and notes that many of his colleagues at Washington University became lifelong friends. Olson feels that any of his accomplishments can be traced to his experiences there.

Since leaving the Glaser laboratory, Olson has had a most distinguished career. After spending eleven years in the Department of Biochemistry and Molecular Biology of the M. D. Anderson Cancer Center in Houston, where he rose to become chairman, in 1995, he was appointed chairman of the Department of Molecular Biology at the U. T. Southwestern Medical Center in Dallas. There, he has held a number of named chairs and professorships. His research has focused on the transcriptional regulation of the development of skeletal muscle, cardiac muscle, and vascular smooth muscle, but the work has encompassed many other organ and tissue systems. Much of his work has involved studies of the myogenic proteins MyoD, myogenin, and HANDs, which are members of the loop-helix-loop (HLH) family of proteins that bind to regulatory sequences in DNA to control the transcription of muscle proteins. Another regulatory protein he studied was the myocyte enhancer protein (MEF-2), which can act in conjunction with MyoD to regulate the expression of genes that give muscle cells their mature state. This transcription factor plays a role in the development of cardiac, skeletal, and vascular smooth muscle, but is also involved in memory and the development of bone and cartilage. In addition to MEF2, several HAND proteins were found to be needed for vascular development/cardiac ventricle formation.

Olson's research and interests have been so widespread that it is difficult to do them justice and to provide adequate citations. For example, he has explored the roles of the calcium-dependent phosphatase calcineurin and the calmodulin-dependent protein kinase CamKII in the control of cardiac growth and function. He has also studied histone deacetylases, which regulate chromatin stability and play a role in cardiac growth, heart failure, and brain development. Another project involves myocardin, which is a master regulator of gene expression in smooth muscle. His more recent work has involved the regulation of cardiomyocyte proliferation by microRNAs. He has received numerous awards, including the Edgar Haber Award and the Basic Research Prize from the American Heart Association. He was elected a fellow of the Heart Association and received their Founding Distinguished Scientist Award. He also received the Pasarow Foundation Award in Cardiovascular Medicine. He was elected a fellow of the American Academy of Arts and Sciences in 1998 and was elected to the National Academy of Sciences and to the Institute of Medicine in 2000. He has also received other prestigious awards, including election as fellow of the American Academy of Microbiology.

Research with Jeffrey Gordon

Another noted scientist who worked with Glaser is Jeffrey Gordon, although he was not formally one of his students. Gordon obtained his bachelor's degree from Oberlin College and his MD from the University of Chicago in 1973. After internship and residency at Barnes Hospital in St. Louis, he joined the Laboratory of Biochemistry at the National Cancer Institute in 1975. He returned to Barnes Hospital in 1978 to become senior assistant resident and then chief resident in medicine. He then rose through the ranks at Washington University to become professor of Medicine and Biological Chemistry in 1987 and then became head of the Department of Molecular Biology and Pharmacology. He is currently director of the Center for Genome Sciences.

His initial research at Washington University dealt with the structure and synthesis of certain apolipoproteins, which are the core proteins of the molecules that transfer lipids throughout the body. There were also studies of the genes and structures of fatty acid binding proteins. But his major focus was on protein N-myristoylation, in which he made pivotal observations. This is an important co-translational modification of proteins in which they are covalently linked to myristic acid, with consequent modification of their functions. Its importance is illustrated by the fact that 5 percent of all eukaryotic proteins are covalently linked to this 14 carbon fatty acid. This modification increases their lipophilicity and allows them to become transiently associated with cellular membranes and other proteins. Gordon's initial work in this area was done in collaboration with Luis Glaser and was directed to the purification and characterization of the enzymes responsible for the myristoylation, namely myristoyl CoA:protein N-myristoyltransferases. He did extensive studies on the enzymes, including substrate specificity, kinetic and biophysical analyses of its mechanism, effects of its knockout in yeast, and definition of its structure.

Another area of interest was the elucidation of mechanisms involved in the development of gut epithelium, including the characterization of specified cell populations, such as multipotent stem cells. More recently, Gordon has studied the microbial ecology of the gut and has demonstrated the ability of components of the microbiota to induce specific responses in the intestinal epithelium. His current focus is on the mutualistic interactions occurring between humans and the trillions of microbes that colonize their gastrointestinal tract. For example, he is interested in how humans acquire their gut microbiomes, how much they differ between individuals, how they change as a function of development, diet, lifestyle, and culture, and how they contribute to personal variations and disease susceptibility. An important observation from a medical point of view is the demonstration that the gut microbiota of fat mice are different from those of lean mice and have an enhanced capacity for harvesting energy from the diet. Importantly, he found that when the lean mice were given certain gut bacteria, they became fat. A study in obese humans on different diets indicates that the same principle operates. Gordon's findings have sparked headlines and have raised the speculation that transplanting gut bacteria could provide a novel therapeutic approach to the treatment of obesity He and others have obtained evidence that alterations in the intestinal microbiota are linked to such conditions as diabetes, pregnancy, asthma, intestinal bowel diease and other conditions.

Gordon has received much recognition for his contributions, including awards from the American Federation for Clinical Research and the American Gastroenterology Association. He as been elected to the National Academy of Sciences, the Institute of Medicine of the Academy, the American Academy of Arts and Sciences, and as a fellow of the American Association for the Advancement of Science and of the American Academy of Microbiology.

Activities at Washington University and University of Miami

Reverting back to Glaser's career, after receiving his PhD in 1956, he wanted to stay in the United States. Carl Cori appointed him as a lecturer in the department, with the expectation that he would leave within a year or two. However, he remained on the faculty for thirty years! He rose through the ranks and was appointed chairman during the last ten years of his tenure (1975–1986). During his time at Washington University, Glaser engaged in a variety of research projects reflecting his wide-ranging intellectual curiosity. He published extensively, with a focus on the enzymes involved in the synthesis of teichoic acids, which are complex molecules attached to the outer walls of certain bacteria. He also worked with Ernst Helmreich and on the enzymology of phosphorylase and studied the assembly and growth of bacterial cell walls. And he worked on defining the surface components involved in cell adhesion, especially neuronal cell adhesion and the interaction between neurites and Schwann cells. The latter cells insulate nerve axons by forming the myelin sheath. The studies illustrated the role of the heparin-binding domain of the neuronal cell adhesion molecule (NCAM) in these interactions.

In keeping with his wide-ranging interests, Glaser also studied the mechanism of 6-deoxyhexose synthesis, the binding and cellular effects of epidermal growth factor (EGF), and the regulation of Na^+/H^+ exchange by growth factors in various cells. As noted above, he conducted much of his research in association with some exceptionally gifted researchers who were attracted by the power of his intellect and his open style of mentorship. He gave up the chair of biological chemistry at Washington University in 1986 and moved to the University of Miami, where he was appointed professor of biology and biochemistry and also executive vice president and provost. In 2005, he became special assistant to the president of the university and interim dean of the School of Education. Since 2009, he has been interim director of the Master of Arts in International Administration Program. These various positions again illustrate the extent of his intellect. He has served on the editorial boards and as associate editor of some prestigious journals and on several review panels of NIH, the American Heart Association, and the American Cancer Society. Glaser was honored for his services to the University of Miami and named the 2007 Outstanding American by Choice by the U.S. Citizenship and Immigration Service.

In 2005, Representative Ros-Lehtinen, Republican of Florida, paid a special tribute to him in the House of Representatives. She noted that he was a Jewish refugee who had fled Austria at the dawn of the Holocaust and that he understands the experience of many refugees who made the University of Miami the international center that it is. She further noted that his sensitivity and insight had allowed him to fully engage in the academic life of the university and to maintain direct personal contact with its students.

Glaser has remained committed to both research and teaching, and when he resigned as provost in 2005, it was so that he could devote more time to teaching. He describes his philosophy as first, "Never do things halfway, but commit and give all you've got," and second, "Strive for excellence." When he retired as provost and executive vice president, his colleagues stated: "Glaser is absolutely wonderful and totally devoted to the University. He is brilliant and cares very deeply about the school." "Glaser's legacy will be a university that is better, both academically and financially, as a result of his countless contributions and tireless efforts to make this university better." Glaser's contributions at both Washington University and the University of Miami are a striking example of the benefits to the American Academy of Science of those driven from Europe by the vicious policies of the Nazis.

References

Black, B. L., and E. N. Olson. 1998. "Transcriptional Control of Muscle Development by Myocyte Enhancer Factor-2 (MEF2) Proteins." *Ann. Rev. Cell Dev. Biol.* 14: 167–96.
Burger, M. M., and L. Glaser. 1966. "The Synthesis of Teichoic Acids V. Polyglycerolphosphate and Polygalactosylglycerolphosphate." *J. Biol. Chem.* 241: 494–506.
Edmondson, D. G., and E. N. Olson. 1993. "Helix-Loop-Helix Proteins as Regulators of Muscle-Specific Transcription." *J. Biol. Chem.* 268: 755–58.
Glaser, L. 1963. "The Synthesis of Deoxysugars." *Physiol. Rev.* 43: 215–42.
Glaser, L. Personal Communication.
Glaser, L., and D. H. Brown. 1955. "Purification and Properties of Glucose-6-Phosphate Dehydrogenase." *J. Biol. Chem.* 216: 67–79.
Glaser, L., and S. Kornfeld. 1961. "The Enzymatic Synthesis of Thymidine Linked Sugars. II. Thymidine Diphosphate Rhamnose." *J. Biol. Chem.* 236: 1795–99.
Gottlieb, D. I., and L. Glaser. 1980. "Cellular Recognition during Neuronal Development." *Ann. Rev. Neurosci.* 3: 303–18.
Hooper, L. V., and J. I. Gordon. 2001. "Commensal Host-Bacterial Relationships in the Gut." *Science* 292: 1115–18.
Kornfeld, S. 2010. *Biographical Sketch*. St. Louis: Washington University.
Kornfeld, S., and L. Glaser. 1961. "The Enzymatic Synthesis of Thymidine Linked Sugars. I. Thymidine Diphosphate Glucose." *J. Biol. Chem.* 236: 1791–94.
Kornfeld, S., and L. Glaser. 1962. "The Enzymatic Synthesis of Thymidine Linked Sugars. V. Thymidine Diphosphate Aminosugars." *J. Biol. Chem.* 237: 3052–59.
Kresge, N., R. D. Simoni, and R. L. Hill. 2008. "N-Myristoyl Transferase Substrate Selection and Catlysis: The Work of Jeffrey I Gordon." *J. Biol. Chem.* 283: e2.
Merlie, J. P., K. Isenberg, B. Carlin, and E. N. Olson. 1984. "Regulation of Synthesis of Acetylcholine Receptors." *Trends Pharmacol. Sci.* 5: 377–79.
Metzger, B., E. Helmreich, and L. Glaser. 1967. "The Mechanism of Activation of Skeletal Muscle Phophorylase by Glycogen." *Proc. Natl. Acad. Sci. U.S.A.* 57: 994–1001.
Olson, E. N. 1990. "MyoD Family, a Paradigm for Development." *Genes & Development* 4: 1454–61.
Olson, E. N. Personal Communication.
Olson, E. N., K. C. Caldwell, J. I. Gordon, and L. Glaser. 1983. "Regulation of Creatine Phosphokinase Expression during Differentiation of BC_3H1 Cells." *J. Biol. Chem.* 258: 2644–52.
Rothenberg, P., D. Cassel, L. Reuss, and L. Glaser. 1983. "Initial Events in the Interaction of Epidermal Growth Factor with Cells." *Prog. Clin. Biol. Res.* 132C: 109–21.

Rothenberg, P., L. Glaser, P. Schlesinger, and D. Cassel. 1983. "Activation of Na^+/H^+ Exchange by Epidermal Growth Factor Elevates Intracellular pH in A431 Cells." *J. Biol. Chem.* 258: 12644–53.

Towler, D. A., S. P. Adams, S. R. Eubanks, D. S. Towery, E. Jackson-Machelski, L. Glaser, and J. I. Gordon. 1987. "Purification and Characterization of Yeast MyristoylCoA: Protein N-Myristoyl Transferase." *Proc. Natl. Acad. Sci. U.S.A.* 84: 2708–12.

Towler, D. A., S. P. Adams, S. R. Eubanks, D. S. Towery, E. Jackson-Machelski, L. Glaser, and J. I. Gordon. 1988. "Protein N-Myristoyl Transferase Activities from Rat Liver and Yeast Possess Overlapping yet Distinct Peptide Substrate Specificities." *J. Biol. Chem.* 263: 1784–90.

Towler, D. A., J. I. Gordon, S. P. Adams, and L. Glaser. 1988. "The Biology and Enzymology of Eukaryotic Protein Acylation." *Ann. Rev. Biochem.* 57: 69–99.

Turnbaugh, P., M. Hamady, T. Yatsunenko, B. L. Cantarel, A. Duncan, R. E. Ley, M. Sogin, W. J. Jones, B. A. Roe, J. P. Affourtit, M. Egholm, B. Henrissat, A. C. Heath, R. Knight, and J. I. Gordon. 2009. "The Core Gut Microbiome in Obese and Lean Twins." *Nature* 457: 480–84.

Turnbaugh, P. J., R. E. Ley, M. A. Mahowald, V. Magrini, E. R. Mardis, and J. I. Gordon. 2006. "Obesity-Associated Gut Microbiome with Increased Capacity for Energy Harvest." *Nature* 444: 1027–31.

Varki, A. 2010. "Introduction of Stuart Kornfeld." *J. Clin. Invest.* 120: 2635–38

25

Ernst Helmreich

Jovial Bavarian

Research in the Cori Laboratory

Ernst Helmreich spent two periods in the Cori laboratory and became a major figure in biochemistry in Germany. He was born in Munich on July 1, 1922, and received a sound education in a traditional high school (gymnasium). He joined the army at age eighteen but never saw action. This was because he developed septicemia following a routine antityphoid vaccination. Helmreich spent most of the year in the hospital but recovered. He was considered unfit for military service and so decided to study medicine at the University of Munich. After graduation, he worked as a physician at the University Medical Clinic. He was encouraged by two physicians there to do some research on digitalis glucosides. He decided to learn chemistry, but most laboratories were in ruins because of the war. He was accepted by Professor Stefan Goldschmidt of the Technical University of Munich. Goldschmidt was one of the few Jewish scientists who returned to Germany after the war. He succeeded the Nobel Prize winner Hans Fischer as professor and took a particular interest in Helmreich's education in chemistry. Helmreich was appointed *"Privat Dozent"* at the university and studied the metabolism of fructose for his *habilitation*. During his time in Munich, he became acquainted with Feodor Lynen, who won the Nobel Prize in 1964 for the discovery of acetyl-CoA and who is mentioned in several places throughout this book. Helmreich wanted to use radioactive compounds in his research, but these were restricted in the American Zone following World War II, and he had to go to Tübingen in the French Zone. Helmreich spent a short time there and became convinced: *"Ich muss nach Amerika"* ("I must go to America"). Lynen recommended him for a position in the Cori laboratory. He was accepted and applied at once to the U.S. National Academy of Sciences for a fellowship to go to there. He obtained the fellowship and went with his wife and first child to St. Louis, where he stayed for fourteen years—first as a postdoctoral fellow in the Department of Biological Chemistry from 1954 to 1956, where he worked with Carl Cori on studies of tissue permeability, namely the distribution of glucose and pentoses between muscle and plasma and the effect of insulin on this. This work finally convinced Cori that insulin enhanced the penetration of sugars into muscle, a concept that he had resisted for many years.

At the end of 1955, Helmreich and his family returned to Germany, where his second child was born, and in 1956 he immigrated to the United States and in 1961 became an American citizen. In 1956, he became an assistant professor in the Department of Medicine at Washington University, where he carried out fundamental and important research with the distinguished immunologist Herman Eisen and with Milton Kern on antibody synthesis and secretion in isolated lymph node cells. This work had quite an impact and was followed by studies of amino acid transport in the cells. These showed that the transport processes were kinetically asymmetric on the inner and outer surfaces. In 1961, he returned to the Cori laboratory, where he was promoted to associate professor, a position he held until 1968. In this context, he worked with Carl Cori and William Danforth (whose contributions are described in chapter 28) on the effects of work and epinephrine on phosphorylase activity and glycolysis in frog muscle. In this work, a novel method to rapidly terminate the reactions in living muscle was introduced, namely, plunging the muscles into isopentane that had been cooled to the temperature of liquid nitrogen. This allowed the effects of activity on phosphorylase to be quantified and established the controlling role of the conversion of phosphorylase b to a in the glycolytic pathway from glycogen to lactic acid.

Helmreich and Cori also looked at the kinetics of the phosphorylase reaction and the effects of pH, temperature, and AMP on the reaction. The effects of AMP spiked Helmreich's interest in the allosteric regulation of enzymes and the associated conformational changes. In 1963, the concept of allosteric regulation had been formulated by Jacques Monod, Jeffries Wyman and Jean Pierre Changeux, and this became a focus of Helmreich's research. This was reinforced when Monod famously declared: *"Allosterisme c'est le deuxième secret de la vie"* ("Allosterism is the second secret of life"). In St. Louis, Helmreich, working in collaboration with Boyd Metzger, Luis Glaser, and Lewis and Jeanine Kastenschmidt, explored the relationship of subunit interactions to the allosteric properties of phosphorylase. In his work with the Kastenschmidts, he used equilibrium binding of substrates and of AMP to study the allosteric transitions of phosphorylase during its activation. This made possible a quantitative analysis of the allosteric activation of the enzyme.

Like other scientists who worked in the Cori department, Helmreich admired Carl Cori as a scientist and as a cultured and erudite person. As expected from his background, Helmreich spoke German and Italian fluently, but what impressed Cori was that Helmreich could recite with perfect recall classical German poetry, including works by Friedrich Schiller, Heinrich Heine, Rainer Rilke, and Johann von Goethe. Helmreich was also influenced by Cori's belief that an enzymologist should find how an enzyme that he characterized in vitro actually functions in the intact cell. But there was a downside to Cori's intellectual magnetism, and this was Helmreich's difficulty in freeing himself from the force of this intellect in order to set up his own laboratory.

Appointment at the University of Würzburg

In 1967, Helmreich went on a sabbatical to Göttingen to work with the legendary Manfred Eigen, who won the Nobel Prize in Chemistry in 1967 for his studies on measuring fast chemical reactions. Helmreich utilized Eigen's method of using rapid changes in temperature to study allosteric changes in phosphorylase, but the approach didn't work, because the enzyme underwent several transitions that were difficult to separate in time. During his time in Germany, Helmreich's friends and colleagues tried to persuade him to return to Bavaria because of his love for hiking and skiing in the Alps. An opportunity arose in 1968 after Carl Martius, the chairman of the Department of Physiological Chemistry at the University of Würzburg, moved to the ETH (Eidgenössische Technische Hochschule), the renowned research institute in Zurich. Martius' successor at the University of Würzburg, Fritz Turba, died a short time later. Helmreich received the call to become the chair at this prestigious university but had difficulty making up his mind and vacillated several times. Würzburg is in an attractive valley of the Main River, with excellent wine-growing areas and is not too far from the Alps.

The university was founded in 1402 but closed in 1413, because the rector was assassinated by his manservant. It reopened in 1575 on a more permanent basis. Its reputation was enhanced by the fact that many notable scientists had been on the faculty, including Rudolf Virchow, Albert von Koelliker, Theodor Boveri, Julius von Sachs, and the Nobel Laureates Emil Fischer, Wilhelm Wien, Wilhelm Röntgen, Hartmut Michel, and Eduard Buchner. Helmreich's final decision to accept the position coincided with a talk by Manfred Eigen in Munich just after he had learned that he had won the Nobel Prize. At a celebration with Eigen at a Munich bar, Helmreich sent a postcard to his wife in St. Louis informing her of his decision. His family, who had been in St. Louis for fourteen years, did not receive the news with enthusiasm, particularly since the decision had been made during a celebration in a bar! Helmreich did not regret taking the position and returning to his homeland. But the change from the "mecca" of modern biochemistry at St. Louis straight to a German university with its teaching responsibilities and bureaucracy was not easy.

Research on Phosphorylase

At Würzburg, Helmreich researched the regulation of metabolism, the central theme of the department. His research focused initially on how pyridoxal phosphate functions in phosphorylase, but Carl Cori felt he should work on adenylyl cyclase and how it was activated by β-adrenergic stimulation. Parenthetically, Helmreich later worked on the regulation of the cyclase. Cori was not enthusiastic about the pyridoxal phosphate project since he considered that pyridoxal phosphate was already established as a cofactor for all enzymes involved in amino acid metabolism. But Helmreich knew from work in the Cori laboratory that pyridoxal phosphate was a stoichiometric constituent of muscle phosphorylase. Furthermore, Edmond Fischer had shown that pyridoxal phosphate was absolutely required for activity of the enzyme, and he and Edwin Krebs had also found that phosphorylase treated with sodium borohydride to irreversibly attach the cofactor retained enzymatic activity. This showed that the role played by pyridoxal phosphate was different from that for other enzymes utilizing this cofactor. Helmreich and his colleagues in Würzburg used physical methods such as ^{31}P-NMR to show that in all active forms of phosphorylase, the 5′-phosphate of pyridoxal phosphate was dianionic. This provided an explanation for the essential role of the cofactor—it functioned as a proton-donor-acceptor shuttle in general acid-base catalysis. Neil Madsen, a former postdoctoral student in the Cori laboratory, had a different interpretation, namely, that there was a direct interaction between the phosphorus of pyridoxal phosphate and the oxygen of the substrate phosphate This disagreement was not resolved until the reaction of phosphorylase with glycosylic substrates was studied by Dieter Palm and Helmut Klein in Helmreich's laboratory. This confirmed a general acid-type activation mechanism and gave considerable support for the phosphate group of pyridoxal phosphate in a proton shuttle.

It gave Helmreich great satisfaction that his proposed catalytic mechanism was shown to be compatible with the three-dimensional structure of phosphorylase deduced from X-ray crystallography by Louise Johnson at Oxford University. Apart from the proof of the catalytic mechanism, the work of X-ray crystallographers has revealed many new and interesting aspects, including the control of activity by allosteric activation. All of this work has made phosphorylase one of the best-studied allosteric enzymes. Helmreich considers that it deserves a place alongside hemoglobin, which, due to the work of Max Perutz (Nobel Prize 1962), is the allosterically regulated protein "par excellence."

Studies on G Proteins

In addition to studying the mechanisms of the regulation of phosphorylase, Helmreich and his associates worked on the regulation of adenylyl cyclase by guanine nucleotides and β-adrenergic agonists. This was pioneering work and principally involved Thomas Pfeuffer. It showed that stable analogs of GTP markedly stimulated the adenylyl cyclase activity of pigeon erythrocyte membranes incubated with isoproterenol, and synergism was observed between the β-agonist and the GTP analog. The effect was also seen in enzyme preparations solubilized from the membranes. A GTP analog was also found to bind to the membranes with high affinity. An important finding was that chromatography of the activated adenylate cyclase solubilized from membranes resulted in separation of a fraction that bound the GTP analog from fractions containing most of the cyclase activity. This was an important advance since it showed that adenylyl cyclase and the GTP-binding protein were separate entities. Pfeuffer subsequently purified the GTP-binding protein and showed that the activation of the cyclase by isoproterenol or fluoride was transmitted by it. The protein was also shown to be labeled by ^{32}P-NAD and to be a target of cholera toxin, a known activator of adenylyl cyclase. Pfeuffer independently engaged in the purification of adenylyl cyclase and succeeded, using a forskolin affinity column—this was a clever approach since forskolin is an activator of the enzyme. When β-adrenergic receptors were extracted from the membranes using detergents and purified by affinity chromatography, they were shown to couple functionally to G_s, the stimulatory G protein for adenylyl cyclase. The receptor was shown to act catalytically in the reconstituted system just as it does in native membranes. In a combined effort by their respective laboratories, Helmreich, Pfeuffer, and Alex Levitzki from Israel finally succeeded, utilizing purified preparations of the receptor, G protein, and cyclase to reconstitute the complete signaling system in lipid vesicles. This work was published only a few weeks after Alfred Gilman and his laboratory in Dallas had also reported the reconstitution of the system. In later work, the $β_2$-adrenergic receptor was produced in baculovirus-infected insect cells and shown to bind a β-adrenergic ligand and to activate G_s. Further work with β-adrenergic receptors purified from erythrocytes showed that these could interact with G protein βγ subunits. Furthermore, the subunits promoted the interaction of the receptor with G_s and increased its GTPase activity.

Helmreich and his colleague Guido Hartmann developed a plan for establishing a biocenter at Würzburg, which would bring together research in biology, chemistry, and medicine under one roof and would incorporate the latest in technology and laboratory design. Carl Cori was not impressed that Helmreich would miss an important scientific meeting in order to seek funds for the biocenter. He didn't see the need for new laboratories since those built for Rudolf Virchow in 1848 were adequate, in his opinion! Cori said to him: *"Es ist nicht wichtig, dass der Käfig aus Gold ist, die Hauptsache ist, dass der vogel singt"* ("It is not important that the cage is golden; all that counts is that the bird sings"). Not all the Würzburg faculty were enthusiastic about the golden cage, but it was eventually built and inaugurated in 1992. (Figure 22).

Helmreich returned to the Department of Biological Chemistry of Washington University in 1981–1982 to work in the laboratory of Elliot Elson, and, together with Yoav Henis, they showed by fluorescence recovery after photobleaching that most of the antagonist-bound β-adrenergic receptor is immobilized in the membranes of liver cells. This suggested that the G protein is sequestered in the same local regions as the receptor. This work convinced him of the need for better fluorescent probes, and in 1993, he spent time at the Max Planck Institute of Biophysical Chemistry in Göttingen in Thomas Jovin's laboratory, searching for better fluorescent probes of β-adrenergic receptors.

Helmreich retired from the chair at Würzburg in 1991, but he continued his research in the Division of Clinical Biochemistry and Pathobiochemistry at the university, which was under the direction of Ulrich Walter. Together with Helmut Heithier, Helmreich utilized FRET (fluorescent energy transfer) to test whether the α-subunit dissociates from the βγ-subunit of G_s upon activation of the β-adrenergic receptor. This was commonly believed to be the case, but they could not demonstrate it. Until today, this remains a hotly debated issue.

From 1993 to 1996, Helmreich spent time at NIH as a Fogarty Scholar in the laboratory of Earl and Thressa Stadtman. At that time, he decided to write a book titled *The Biochemistry of Cell Signalling*. This was a comprehensive and scholarly treatment of the topic, which was dedicated to Carl Cori and published in 2001 by Oxford University Press. The preface begins with a quotation from Goethe: *"Alles Gescheite ist schon gedacht worden, man muss nur versuchen, es noch einmal zu denken"* ("Everything worthwhile and intelligent has already been thought. One can only try to think it over again").

Helmreich has been the recipient of numerous honors. These include presidency of the German Society of Biological Chemistry, the Order of Merit of the Federal Republic of Germany, and election to the prestigious *Leopoldina*, the German Academy of Science, and the Bavarian Academy of Science. The *Leopoldina* awarded him its Cothenius Medal for his scientific achievements, and the University of Würzburg honored him with the Rinecker Medal. He has also served on the Senate of the Deutschen Forschungsgemeinschaft (DFG), which is the principal source of funds for research in Germany, the Council of the International Union of Biochemistry, and the Council of the European Molecular Biology Laboratory.

In 2007, he and his wife Mylein returned to their beloved Bavarian Alps. He continues to write commentaries and reviews.

References

Cassel, D., and T. Pfeuffer. 1978. "Mechanism of Cholera Toxin Action: Covalent Modification of the Guanyl Nucleotide Binding Protein of the Adenylate Cyclase System." *Proc. Natl. Acad. Sci. U.S.A.* 75: 2669–73.

Cassel, D., and Z. Selinger. 1977. "Mechanism of Adenylate Cyclase Activation by Cholera Toxin: Inhibition of GTP Hydrolysis at the Regulatory Site." *Proc. Natl. Acad. Sci. U.S.A.* 74: 3307–11.

Danforth, W. H., E. Helmreich, and C. F. Cori. 1962. "The Effect of Contraction and of Epinephrine on the Phosphorylase Activity of Frog Sartorius Muscle." *Proc. Natl. Acad. Sci. U.S.A.* 48: 1191–99.

Feder, D., M. J. Im, H. W. Klein, M. Hekman, A. Holzhöfer, C. Dees, A. Levitski, E. J. M. Helmreich, and T. Pfeuffer. 1986. "Reconstitution of Beta1 Adrenoceptor-Dependent Adenylate Cyclase from Purified Components." *EMBO J.* 5: 1509–14.

Feldmann, K., and W. E. Hull. 1977. "^{31}P nuclear Magnetic Resonance Studies of Glycogen Phosphorylase from Rabbit Skeletal Muscle: Ionization States of Pyridoxal 5'-Phosphate." *Proc. Natl. Acad. Sci. U.S.A.* 74: 856–60.

Fischer, E. H., A. B. Kent, E. R. Snyder, and E. G. Krebs. 1958. "The Reaction of Sodium Borohydride with Muscle Phosphorylase." *J. Am. Chem. Soc.* 80: 2906–07.

Hekman, M., D. Feder, A. K. Keenan, A. Gal, H. W. Klein, T. Pfeuffer, A. Levitski, and E. J. M. Helmreich. 1984. "Reconstitution of Beta-Adrenergic Receptor with Components of Adenylate Cyclase." *EMBO J.* 3: 3339–45.

Helmreich, E. J. M. 1994. "Excursions in Biophysics by a Classical Biochemist." *Protein Science* 3: 528–32.

Helmreich, E. J. M. 1995. "Recollections: Vacillations of a Classical Biochemist." *Comprehensive Biochemistry* 38: 163–91.

Helmreich, E., and C. F. Cori. 1957. "Studies of Tissue Permeability. II. The Distribution of Pentoses between Muscle and Plasma." *J. Biol. Chem.* 224: 663–80.

Helmreich, E., and C. F. Cori. 1964. "The Role of Adenylic Acid in the Activation of Phosphorylase." *Proc. Natl. Acad. Sci. U.S.A.* 51: 131–38.

Helmreich, E., M. Kern, and H. N. Eisen. 1961. "The Secretion of Antibody by Isolated Lymph Node Cells." *J. Biol. Chem.* 236: 464–73.

Helmreich, E., and D. M. Kipnis. 1962. "Amino Acid Transport in Lymph Node Cells." *J. Biol. Chem.* 237: 2582–89.

Henis, Y. I., M. Hekman, E. L. Elson, and E. J. M. Helmreich. 1982. "Lateral Motion of Beta Receptors in Membranes of Cultured Liver Cells." *Proc. Natl. Acad. Sci. U.S.A.* 79: 2907–11.

Johnson, L. N., K. R. Acharya, M. D. Jordan, and P. J. McLaughlin. 1990. "Refined Crystal Structure of the Phosphorylase-Heptulose 2-Phosphate-Oligosaccharide-AMP Complex." *J. Mol. Biol.* 211: 645–61.

Karpatkin, S., E. Helmreich, and C. F. Cori. 1964. "Regulation of Glycolysis in Muscle. II. Effect of Stimulation and Epinephrine on Isolated Frog Sartorius Muscle." *J. Biol. Chem.* 239: 3139–45.

Kastenschmidt, L. L., J. Kastenschmidt, and E. Helmreich. 1968. "Subunit Interactions and Their Relationship to the Allosteric Properties of Rabbit Skeletal Muscle Phosphorylase *b*." *Biochemistry* 7: 3590–3608.

Kastenschmidt, L. L., J. Kastenschmidt, and E. Helmreich. 1968. "Effect of Temperature on the Allosteric Transitions of Rabbit Skeletal Muscle Phosphorylase *b*." *Biochemistry* 7: 4543–56.

Klein, H. W., M. J. Im, D. Palm, and E. J. M. Helmreich. 1984. "Does Pyridoxal 5'-Phosphate Function in Glycogen Phosphorylase as an Electrophilic or a General Acid Catalyst?" *Biochemistry* 23: 5853–5961.

Metzger, B., E. Helmreich, and L. Glaser. 1967. "The Mechanism of Activation of Phosphorylase by Glycogen." *Proc. Natl. Acad. Sci. U.S.A.* 57: 994–1001.

Pfeuffer, T. 1977. "GTP-Binding Proteins in Membranes and the Control of Adenylate Cyclase Activation." *J. Biol. Chem.* 252: 7224–34.

Pfeuffer, T., and E. J. M. Helmreich. 1975. "Activation of Pigeon Erythrocyte Membrane Adenylate Cyclase by Gyanylnucleotide Analogues and Separation of a Nucleotide Binding Protein." *J. Biol. Chem.* 250: 867–76.

Reiländer, H., F. Boege, S. Vasudevan, G. Maul, M. Hekman, C. Dees, W. Hampe, E. J. M. Helmreich, and H. Michel. 1991. "Purification and Functional Characterization of the Human Beta 2-Adrenergic Receptor Produced in Baculovirus-Infected Insect Cells." *FEBS Lett.* 282: 441–44.

Withers, S. G., N. B. Madsen, B. D. Sykes, M. Takagi, S. Shimomura, and T. Fukui. 1981. "Evidence for Direct Phosphate-Phosphate Interaction between Pyridoxal Phosphate and Substrate in the Glycogen Phosphorylase Catalytic Mechanism." *J. Biol. Chem.* 256: 10759–62.

26

Carl Frieden

Enzyme Kineticist

In 1955, Carl Frieden moved to the Cori department to be a postdoctoral fellow with Sidney Velick (Figure 23). Frieden was born in New Rochelle, New York, in 1928 and graduated from Carleton College in Minnesota in 1951. He started graduate studies with Karl Link in the Biochemistry Department at the University of Wisconsin, Madison. Link was a famous but unorthodox carbohydrate chemist. He had discovered the active ingredient from spoiled sweet clover hay that caused a bleeding disorder in cattle because it stopped the blood from clotting. The hemorrhagic factor was isolated and characterized by Link and named dicumarol. This was synthesized and shown to be identical to the natural product. Its structure was recognized to be similar to vitamin K, and its anticlotting action is due to its inhibition of the synthesis of vitamin K-dependent clotting factors. A derivative called warfarin (Coumadin) is now widely used as a rodenticide and also an anticoagulant in many human diseases. The patent on this has brought hundreds of millions of dollars to the University of Wisconsin. Link loved to tell the story of how a farmer came to the biochemistry building during a blizzard carrying a milk can of blood from the affected cows and one hundred pounds of spoiled clover. The farmer had traveled almost a hundred miles to Madison, and, by pure chance, he came to the university because the office of the State Veterinarian was closed. Link enjoyed recounting the incident and also the earthy guttural comments of his assistant, who punctuated them with Schwabian German. Link was an eccentric individual with flowing locks, extreme liberal political views, and an eagerness to defend unpopular causes. He generally adopted an unusual attire (large bow ties, work shoes, knickers, and sometimes a cape). He was an effective lecturer but sometimes injected his lectures with outrageous statements, slurs against the university administration, and pithy quotations. He had a tendency (an eagerness) to argue with other faculty members, and his quick temper drove him to throttle in public one of the scientists whom he particularly disliked!

Frieden decided that he wasn't interested in nutrition but preferred protein (enzyme) structure. Link then contacted Robert (Bob) Alberty in the Chemistry

Department, and Alberty accepted Frieden as a graduate student. Alberty was an expert on the thermodynamics of enzymatic reactions, and Frieden was assigned to study the kinetics of fumarase, an enzyme in the citric acid cycle. The enzyme was purified from pig hearts, necessitating many trips to the Oscar Meyer factory in Madison. Purification and crystallization of the enzyme were rather fractious, but Frieden did obtain crystals, although he was disappointed to learn that another investigator had also crystallized the enzyme.

Studies on Glutamate Dehydrogenase

Frieden described Alberty as a wonderful mentor who guided him into the area of enzyme kinetics, which at that time was not well advanced. When Frieden moved to Washington University for his postdoctoral with Sidney Velick, he was left virtually alone and published three sole-authored papers! After two years, Carl Cori hired him and asked him what he wanted to work on. This surprised Frieden, who expected Cori to signify the project he wanted him to work on. He searched the literature for an interesting enzyme and settled on liver glutamate dehydrogenase. This was of particular interest because it had three substrates (α-ketoglutarate, ammonia, and either NADH or NADPH). With enzymes with multiple substrates, an issue is what happens if the substrates are added sequentially or randomly. Frieden found that the kinetic equations were different, depending on how the substrates were added and concluded that the order was sequential.

Frieden searched for some competitive inhibitors of the enzyme utilizing initial velocity measurements. He looked at moeities of either NAD or NADP and found an effect of ADP, but this stimulated the reaction instead of producing the expected inhibition. He concluded that the nucleotide was binding at a site different from the catalytic site. This observation initiated studies of the effects of various adenine and guanine nucleotides, which revealed that the nucleotides either activated or inhibited, depending on the conditions. After the results were published in the *Journal of Biological Chemistry*, Jacques Monod from the Institut Pasteur made a brief visit to Washington University and invited Frieden to spend a sabbatical in his laboratory in Paris. His visit was very short, but at a time when history was being made. This was when Jean-Pierre Changeux was reinvestigating the finding that threonine deaminase did not obey Michaelis-Menten kinetics. Frieden spent a lot of time deriving equations to fit the data—all to no avail. But the findings eventually led to the concept of allosteric enzymes and the development of the brilliant theory of allostery by Monod, Jeffries Wyman, and Changeux, published in 1965.

Although aspartate transcarbamylase was the original model for an allosteric enzyme, Frieden realized glutamate dehydrogenase also filled the bill, because ligands that bound distant from the active site affected its activity. He continued looking at the effects of NADH and NADPH and found that high concentrations of NADH were inhibitory, whereas NADPH was not. In further experiments, Dave Bates, a graduate student, found that the inhibitory effect of NADH was time-dependent. This caused Frieden to think about initial enzyme velocities—if such a velocity is time-dependent, what does it mean? This led to the concept of hysteresis, that is, when the change in activity lags behind the initial formation of the enzyme-substrate complex, reflecting a time-dependent conformational change. Frieden developed an equation for this hysteretic behavior, which became useful to others.

He then focused on how to analyze time-dependent processes, but the derivation of equations to describe the full-time course of an enzymatic reaction was found to be a horrendous experience and was worse with multisubstrate enzymes. However, Dave Bates tackled the problem and wrote a program for a mainframe computer. Bruce Barshop, an MD, PhD student, then made the program more user-friendly, and, as KinSim, it is still in use and has been adapted for desktop computers. Frieden then applied KinSim to an analysis of the kinetics of bacterial dihydrofolate reductase, which has a complex kinetic mechanism involving substrate-product complexes with almost no free enzyme produced in the reaction.

Research with Roberta Colman

Frieden found that glutamate dehydrogenase had the interesting property of polymerization, and, in collaboration with Roberta Colman, his first postdoctoral student, he studied the differences in the binding of different purine nucleotides to the monomeric and polymeric forms. Colman also studied the effects of acetylation of the enzyme on its catalytic and regulatory properties and concluded that this modification caused a loss of communication between the coenzyme binding site and the nucleotide binding site. From these studies, it was concluded that there was a relationship between the functional group acetylated and a rate-limiting step in the enzymatic reaction, and that this group was also responsible for the interaction between the coenzyme and nucleotide site in the active enzyme. In further work, the effects of acetylation on the monomer-polymer state of the enzyme were examined. Although low concentrations of acetic anhydride decreased enzymatic activity drastically, there was little change in the molecular weight of the enzyme. At higher concentrations of the reagent, the molecular weight dropped sharply to that of the monomer, and activity was completely lost. Thus, Frieden and Colman concluded that different functional groups were responsible for catalytic activity and monomer-tetramer association.

Roberta Colman went on to a very impressive career in biochemistry. She continued to look at the effects of chemical modifications on enzymes and also developed specific affinity labels for regulatory sites. In addition to glutamate dehydrogenase, other enzymes she studied from the point of view of relating structure to function included isocitrate dehydrogenases, adenylosuccinate lyase, and glutathione

S-transferase. In 1973 she was appointed to a named chair of chemistry and biochemistry at the University of Delaware. She won numerous awards, including the Herbert A. Sober Award of the American Society for Biochemistry and Molecular Biology (ASBMB) and was elected a fellow of the American Association for the Advancement of Science. She served in many leadership positions in biochemical societies and on many review and editorial boards.

When the structure of bovine liver glutamate dehydrogenase was solved by another group, Frieden found excellent agreement between his identification of amino acid residues related to allosteric properties—using chemical modification—and the final structure. As noted above, the mammalian enzyme uses both NADH and NADPH, and it uses them equally well. On the other hand, the enzymes from prokaryotes do not show this behavior. With Barry Goldin, Frieden speculated that when the enzyme used NAD or NADH, it functioned in a catabolic role, whereas when it used NADP or NADPH, it functioned in an anabolic role.

Frieden has been involved in other projects besides glutamate dehydrogenase. These include the kinetics of other regulatory enzymes, such as phosphofructokinase, and the process of actin polymerization. He has also investigated the mechanism of protein folding utilizing several different proteins: intestinal fatty acid binding protein, adenosine deaminase, dehydrofolate reductase, and a protein required for the formation of pili (filamentous appendages) in bacteria. These proteins were chosen for specific reasons, and the folding was explored by site-directed mutagenesis, ^{19}F and proton NMR, circular dichroism, fluorescence, and X-ray crystallography. In addition, the role of chaperones (proteins that assist folding) was examined in certain instances. The dynamics and aggregation properties of intrinsically disordered proteins similar to those involved in certain neurological diseases (Alzheimer's and Huntington's) and the association/disassociation of apolipoproteins are also receiving attention.

During his long and distinguished career, Frieden has received many awards. He was previously Alumni Endowed Professor of Biochemistry and Molecular Biophysics at Washington University (1994–2000) and chair of the Department (2000–2005). He was elected to the National Academy of Sciences in 1988 and to the American Academy of Arts and Sciences in 2004. He was elected as a fellow of the American Association for the Advancement of Science in 1988 and received the Christian Anfinsen Award from the Protein Society in 2007. He has served on NIH Study Sections and on the editorial boards of several journals. He remains active in research at over 80 years of age.

References

Bates, D. J., and C. Frieden. 1973. "A Small Computer System for Routine Analysis of Enzyme Kinetic Mechanisms." *Comput. Biomed. Res.* 6: 474–86.

Burris, R. H. 1994. "Karl Paul Link January 31,1901–November 21, 1978." *Biog. Mem. Natl. Acad. Sci. U.S.A.* 605: 176–95.

Coffee, C. J., R. A. Bradshaw, B. R. Goldin, and C. Frieden. 1971. "Identification of the Sites of Modification of Bovine Liver Glutamate Dehydrogenase Reacted with Trinitrobenzenesulfonate." *Biochemistry* 10: 3516–26.

Colman, R. F., and C. Frieden. 1966. "Cooperative Interaction between the GTP Binding Sites of Glutamate Dehydrogenase." *Biochem. Biophys. Res. Commun.* 22: 100–05.

Colman, R. F., and C. Frieden. 1966. "On the Role of Amino Groups in the Structure and Function of Glutamate Dehydrognase. I. Effect of Acetylation on Catalytic and Regulatory Properties." *J. Biol. Chem.* 241: 3652–60.

Colman, R. F., and C. Frieden. 1966. "On the Role of Amino Groups in the Structure and Function of Glutamate Dehydrogenase. II. Effect of Acetylation on Molecular Properties." *J. Biol. Chem.* 241: 3661–70.

Frieden, C. 1958. "The Dissociation of Glutamate Dehydrogenase by Reduced Diphosphopyridine Nucleotide (DPN)." *Biochim. Biophys. Acta* 27: 431–32.

Frieden, C. 1957. "The Calculation of an Enzyme-Substrate Dissociation Constant from the Overall Initial Velocity for Reactions Involving Two Substrates." *J. Am. Chem. Soc.* 79: 1894–96.

Frieden, C. 1958. "The Influence of the Ionization of a Group in the Substrate Molecule on the Kinetic Parameters of Enzyme Reactions." *J. Am. Chem. Soc.* 80: 6519–23.

Frieden, C. 1959. "Glutamic Dehydrogenase. II. The Effect of Various Nucleotides on the Association-Dissociation and Kinetic Properties." *J. Biol. Chem.* 234: 815–20.

Frieden, C. 1959. "Glutamic Dehydrogenase. III. The Order of Substrate Addition in the Enzymatic Reaction." *J. Biol. Chem.* 234: 2891–96.

Frieden, C. 1962. "The Unusual Inhibition of Glutamate Dehydrogenase by Guanosine Di- and Triphosphate." *Biochim. Biophys. Acta* 59: 484–86.

Frieden, C. 1970. "Kinetic Aspects of Regulation of Metabolic Processes. The Hysteretic Enzyme Concept." *J. Biol. Chem.* 245: 5788–99.

Frieden, C. 2008. "A Lifetime of Kinetics." *J. Biol. Chem.* 283: 19873–78.

Frieden, C. Personal Communication.

Goldin, B. R., and C. Frieden. 1971. "Effect of Trinitrophenylation of Specific Lysyl Residues on the Catalytic, Regulatory, and Molecular Properties of Bovine Liver Glutamate Dehydrogenase." *Biochemistry* 10: 3527–34.

Goldin, B. R., and C. Frieden. 1972. "The Effect of Pyridoxal Phosphate Modification on the Catalytic and Regulatory Properties of Bovine Liver Glutamate Deydrogenase." *J. Biol. Chem.* 247: 2139–44.

Kresge, N., R. D. Simoni, and R. L. Hill. 2005. "Hemorrhagic Sweet Clover Disease Dicumarol, and Warfarin: The Work of Karl Paul Link." *J. Biol. Chem.* 280: e5.

Monod, J., J. Wyman, and J.-P. Changeux. 1965. "On the Nature of Allosteric Transitions: A Plausible Model." *J. Mol. Biol.* 12: 88–118.

Smith, T. J., P. E. Peterson, T. Schmidt, J. Fang, and C. A. Stanley. 2001. "Structures of Bovine Glutamate Dehydrogenase Complexes Elucidate the Mechanism of Purine Regulation." *Mol. Biol.* 307: 707–20.

27

David Kipnis

Focus on Diabetes

Studies on Muscle Carbohydrate Metabolism

David Kipnis was born and raised in Baltimore and received his MD from the University of Maryland. He trained in medicine at Johns Hopkins, Duke, and the University of Maryland. He then joined the Cori laboratory in 1955 after meeting the Coris in Atlantic City (Figure 24). He had also applied to work with Fritz Lipmann, but Lipmann was out of the country and did not reply. Kipnis found the atmosphere in the Cori laboratory very stimulating with great discussions in the library where all had lunch. He was assigned to work with Ernst Helmreich whose career was described in chapter 25. Cori gave Kipnis a new project, but according to custom, the topic was related to one previously worked on in the laboratory. He gave Kipnis a folder of papers, and the topic that interested him the most was how insulin worked. Kipnis was taken aback when Cori asked Helmreich to teach him because he had no experience in a lab! As described in chapter 16 on Charles Park, it was known in the diabetes field that Rachmiel Levine had done some whole animal studies that suggested that insulin acted by stimulating the movement of glucose into cells. When Kipnis started studying this in isolated diaphragm muscle, he was greatly frustrated by his failure to see an insulin effect. This was due to the uncontrolled influx of glucose through the cut edges of the diaphragms. Later experiments with intact diaphragms (including the rib cage) gave positive results. These studies utilized pentose sugars, which cannot be metabolized by muscle, and showed that insulin increased their penetration. However, it was important to show this for glucose. Studies of the distribution of free glucose in plasma and muscle water of the diaphragm and gastrocnemius carried out with Helmreich showed that epinephrine caused an intracellular accumulation of glucose and glucose-6-phosphate. However, insulin was without effect unless doses were used that triggered a hypoglycemic response i.e. the release of epinephrine. In further experiments, 2-deoxyglucose was used instead of glucose since this sugar can penetrate into cells and be phosphorylated, but not further metabolized, unlike glucose, which is rapidly phosphorylated and metabolized inside muscle cells. The results showed that insulin caused the intracellular accumulation of 2-deoxyglucose-6-phosphate, but not 2-deoxyglucose. Thus the results did not resolve whether insulin affected glucose transport or phosphorylation.

The experiments were repeated with diabetic and adrenalectomized rats. In the diaphragms of severely diabetic animals, phosphorylation of 2-deoxyglucose was decreased, but there was no accumulation of the free sugar, suggesting impairments in both transport and phosphorylation. Direct addition of insulin to the diaphragms increased the accumulation of 2-deoxyglucose, indicating an effect on penetration. However, insulin did not normalize the phosphorylation, but this was restored to normal by adrenalectomy. It was known, since the work of Bernado Houssay, that the adrenal and pituitary glands could modify carbohydrate metabolism, but the point of action of their hormones was unknown. Thus these results pinpointed glucose phosphorylation in muscle as one of their targets. In all of these studies, the results were interpreted with extraordinary care as shown by kinetic analyses and many footnotes. In other words, the influence of Carl Cori was evident. The findings of Kipnis utilizing rat diaphragms were essentially identical to those of Charles Park, who employed the perfused rat heart (see chapter 16).

Clinical Studies Involving Insulin and Starvation

In 1960, Kipnis moved to the Department of Medicine at Washington University, where he embarked on a combination of clinical and basic studies. In the area of clinical research, he studied insulin and growth hormone secretion in human obesity and acromegaly and also the growth-promoting and anti-insulin effects of growth hormone. These studies were done in collaboration with William Daughaday, whose career was described in chapter 21. Another study described the plasma insulin responses when the same dose of glucose was administered orally or intravenously. Surprisingly, the responses to the intravenous load were less than half those seen with the oral load. This indicated that some factors in the gastrointestinal tract influenced the insulin response. It is now recognized that several gastrointestinal hormones can stimulate insulin secretion. In associated studies, obese individuals showed a greater than normal insulin response to oral glucose, whereas in diabetics, the response was markedly impaired, as expected.

Other clinical studies of significance included the demonstration that the sympathetic system plays a major role in counteracting insulin-induced hypoglycemia because of glycogen breakdown in the liver. At that time, there were few studies of the in vivo role of glucagon, and Kipnis used neutralizing antibodies to this hormone to show it played a role in maintaining blood glucose. In a complementary study, a patient with a pancreatic tumor that over-secreted glucagon was shown to exhibit uncontrolled diabetes.

One of Kipnis' most outstanding clinical studies was to explore the changes in fuel supply and hormones in men starved for seven days. This was one of the classic studies of metabolism carried out in humans. It involved seven other investigators, principally George Cahill Jr., and was an extraordinarily detailed investigation. Cahill was a professor at Harvard Medical School and director of the Elliott P. Joslin Research Laboratory. He was a leading researcher in the field of diabetes and served as the medical director of the Howard Hughes Medical Institute. The investigators found that in the first few days of starvation, the carbohydrate stores were a significant contributor to the fuel mix for the body, and glucose was available for use as a fuel by the brain. With time, the body glycogen stores became depleted and the level of glucose in the blood decreased to reach a plateau by the third day. The blood insulin level showed a similar pattern, and, with the fall in insulin, free fatty acids were released from adipose tissue and became the principal fuel for the body. They were oxidized to produce ketone bodies, which later became a source of fuel for the brain when it became adapted to utilize them. At the early stages of starvation, protein breakdown was significant, as revealed by urinary nitrogen excretion, but later, protein was conserved. Many of the amino acids produced by protein breakdown were converted to glucose by gluconeogenesis, and this became the prime source of this fuel. The breakdown of protein was succeeded by a phase of nitrogen conservation. The findings illustrated how superbly the body adapted its metabolism to starvation in order to maintain the supply of fuel to the various tissues, especially the brain. The results also revealed the primacy of insulin as a regulator of the important metabolic changes.

Studies of Cyclic Nucleotides, Insulin Biosynthesis, and Amino Acids

One of Kipnis' most striking contributions was the development of radioimmunoassays for cyclic AMP and cyclic GMP, with the lead investigator being Alton Steiner. These enabled the nucleotides to be measured in tissues and body fluids and demonstrated their control by various hormones. Their findings in rats and mice were similar to those found in humans by Sutherland's group using a different assay. Thus glucagon and parathyroid hormone increased the urinary excretion of cyclic AMP without affecting cyclic GMP. The studies were complemented in in vitro studies in various tissues, and the results correlated with the known physiological actions of the hormones. The work then focused on the brain, and variations in cyclic AMP levels were found in different regions. Interestingly, the level of cyclic GMP was very high in the cerebellum—the significance of this is still not known.

Another major study was of insulin biosynthesis in pancreatic islets, which was principally conducted by Alan Permutt. It was well known that glucose stimulated insulin release from the islets, but Permutt showed that glucose also stimulated the biosynthesis of insulin. Studies with inhibitors showed that the effect was exerted at both the transcriptional and post-transcriptional levels. Thus glucose was found to cause an increase in the mRNA of proinsulin, the precursor of insulin, and also an increase in the translation of this mRNA. A later detailed analysis showed that the effect of glucose was exerted on the initiation phase of proinsulin synthesis and not on the elongation phase. Alan Permutt stayed at Washington University and continued his work on insulin synthesis. Later, he established himself as a world expert on the genetics of diabetes.

Another important series of experiments was carried out in association with Alan Garber and involved a tiny muscle called the epitrochlearis. This muscle is so small that some investigators doubted its existence! It was used to study the regulation of the release of two gluconeogenic amino acids (alanine and glutamine) from muscle. Surprisingly, the release was not related to glucose metabolism and was not due to the breakdown of a protein rich in these amino acids. On the other hand, it reflected the synthesis of alanine and glutamine from amino acids in muscle that were produced from proteolysis. Further work examined the effects of fasting, diabetes, muscular dystrophy, uremia, and various hormones on the release of the amino acids. In his later career, Garber focused on the treatment of type II diabetes.

In 1973, Kipnis was appointed Busch Professor and chairman of the Department of Medicine at Washington University. He was recognized as an authority in the field of diabetes on the basis of his clinical and basic studies. He continued to be active in diabetes research, with a focus on the mechanisms responsible for the disease, and the regulation of carbohydrate and protein metabolism in vitro and in vivo.Kipnis received many awards for his work. He received the Lilly Award, the Banting Lectureship, and the Charles H. Best Award from the American Diabetes Association, and the Ernst Oppenheimer Award from the Endocrine Society. He was also awarded the Kober Medal from the Association of American Physicians, the Banting Lectureship of the British Diabetes Association, and the Elliott Joslin Medal. In recognition of his contributions at the national level, he was elected to the National Academy of Sciences, the Institute of Medicine of the National Academy, and the American Academy of Arts and Sciences.

References

Beck, P., J. H. Koumans, C. A. Winterling, M. F. Stein, W. H. Daughaday, and D. M. Kipnis. 1964. "Studies of Insulin and Growth Hormone Secretion in Human Obesity." *J. Lab. Clin. Med.* 64: 654–67.

Beck, P., D. S. Schalch, M. L. Parker, D. M. Kipnis, and W. H. Daughaday. 1965. "Correlative Studies of Growth Hormone and Insulin Plasma Concentrations with Metabolic Abnormalities in Acromegaly." *J. Lab. Clin. Med.* 66: 366–79.

Cahill, G. F. Jr., M. G. Herrera, A. P. Morgan, J. S. Soeldner, J. Steinke, P. L. Levy, G. A. Reichard Jr., and D. M. Kipnis. 1966. "Hormone-Fuel Interrelationships during Fasting." *J. Clin. Invest.* 45: 1751–69.

Garber, A. J., I. E. Karl, and D. M. Kipnis. 1976. "Alanine and Glutamine Synthesis and Release from Skeletal Muscle. I. Glycolysis and Amino Acid Release." *J. Biol. Chem.* 251: 826–35.

Garber, A. J., I. E. Karl, and D. M. Kipnis. 1976. "Alanine and Glutamine Synthesis and Release from Skeletal Muscle. II. The Precursor Role of Amino Acids in Alanine and Glutamine Synthesis." *J. Biol. Chem.* 251: 836–43.

Garber, A. J., P. E. Cryer, J. V. Santiago, M. W. Haymond, A. S. Pagliari and D. M. Kipnis. 1976. "The Role of Adrenergic Mechanisms in the Substrate and Hormonal Response to Insulin-Induced Hypoglycemia in Man." *J. Clin. Invest.* 58: 7–15.

Kipnis, D. M. 2006. *Transcript of Oral History by Permission of the Archivist.* St. Louis: Becker Medical Library, Washington University School of Medicine.

Kipnis, D. M. 2009. *Biographical Information.* St. Louis: Washington University School of Medicine.

Kipnis, D. M., and C. F. Cori. 1957. "Studies of Tissue Permeability. III. The Effect of Insulin on Pentose Uptake by the Diaphragm." *J. Biol. Chem.* 224: 681–93.

Kipnis, D. M., and C. F. Cori. 1959. "Studies of Tissue Permeability. V. The Penetration and Phosphorylation of 2-Deoxyglucose in the Rat Diaphragm." *J. Biol. Chem.* 234: 171–77.

Kipnis, D. M., and C. F. Cori. 1960. "Studies of Tissue Permeability.VI. The Penetration and Phosphorylation of 2-Deoxyglucose in the Diaphragm of Diabetic Rats." *J. Biol. Chem.* 235: 3070–75.

Kipnis, D. M., E. Helmreich, and C. F. Cori. 1959. "Studies of Tissue Permeability. IV. The Distribution of Glucose between Plasma and Muscle." *J. Biol. Chem.* 234: 165–70.

Leichter, S. B., A. S. Pagliara, M. H. Grieder, S. Pohl, J. Rosai, and D. M. Kipnis. 1975. "Uncontrolled Diabetes Mellitus and Hyperglucagonemia Associated with an Islet Cell Carcinoma." *Am. J. Med.* 58: 285–93.

Perley, M., and D. M. Kipnis. 1966. "Plasma Insulin Responses to Glucose and Tolbutamide of Normal Weight and Obese Diabetic and Nondiabetic Subjects." *Diabetes* 15: 867–68.

Perley, M. J., and D. M. Kipnis. 1967. "Plasma Insulin Responses to Oral and Intravenous Glucose: Studies in Normal and Diabetic Subjects." *J. Clin. Invest.* 46: 1954–62.

Permutt, M. A. 1974. "The Effect of Glucose on Initiation and Elongation Rates in Isolated Rat Pancreatic Islets." *J. Biol. Chem.* 248: 2738–42.

Permutt, M. A., and D. M. Kipnis. 1972. "Insulin Biosynthesis. I. On the Mechanism of Glucose Stimulation." *J. Biol. Chem.* 247: 1194–99.

Steiner, A. L., J. A. Ferrendelli, and D. M. Kipnis. 1972. "Radioimmunoassay for Cyclic Nucleotides. III. Effect of Ischemia, Changes during Development and Regional Distribution of Adenosine 3',5'-Monophosphate and Guanosine 3',5'-Monophosphate in Mouse Brain." *J. Biol. Chem.* 247: 1121–24.

Steiner, A. L., A. S. Pagliari, L. R. Chase, and D. M. Kipnis. 1972. "Radioimmunoassay for Cyclic Nucleotides. II. Adenosine 3',5'-Monophosphate and Guanosine 3',5'-Monophosphate in Mammalian Tissues and Body Fluids." *J. Biol. Chem.* 247: 1114–20.

Steiner, A. L., C. W. Parker, and D. M. Kipnis. 1972. "Radioimmunoassay for Cyclic Nucleotides. I. Preparation of Antibodies and Iodinated Cyclic Nucleotides." *J. Biol. Chem.* 247: 1106–13.

28

William Danforth

Academic Leader

Training in Cardiology and Research in the Cori Laboratory

William Henry Danforth was a postdoctoral fellow in the Cori department from 1961 to 1963 and later rose to great prominence as chancellor of Washington University. He was the grandson of the first William H. Danforth, who graduated from Washington University and founded the Ralston-Purina Company, which is the world's largest supplier of food for dogs, cats, and experimental animals and which is packaged in the familiar trademark checkerboard bags. The first William Danforth and his wife also established the Danforth Foundation as a national educational philanthropy. The present William Danforth continues this philanthropy, as does his brother John, an ordained Episcopal minister and former U.S. senator and ambassador to the United Nations.

Danforth was born in St. Louis on April 10, 1926, and served in the U.S. Navy from 1944 to 1945. He received his BA from Princeton in 1947 and his MD from Harvard in 1951. He was an intern in medicine at the Barnes Hospital in St. Louis during 1951–1952, followed by a stint as a physician in the U.S. Navy during the Korean War. In 1954, he was appointed an instructor in medicine at Washington University and later resident in medicine at the Barnes Hospital. During 1957–1958, he was a fellow in cardiology (Figure 25), when he worked with the notable cardiologist Richard Bing studying the response of the heart to anoxia and ischemia, with a focus on glycolytic reactions and ATP levels. Bing was a native of Nuremberg, Germany, who had emigrated to the United States in 1936 after Hitler rose to power. His first position was at the College of Physicians and Surgeons at Columbia, where he studied under famous surgeon Allen Whipple, who had him explore the vasopressor peptides released by the kidneys. This research area is covered in chapter 6 on Luis Leloir. Whipple must have been impressed with Bing, because he allowed him to marry his daughter! Bing gained prominence at Johns Hopkins, where he was known for establishing, with Alfred Blalock and Helen Taussig, the first cardiac catheterization laboratory dedicated to congenital cardiac disease. Bing was also known for his passionate interest in music as well as medicine, having studied piano and composition at the Nuremberg Conservatory. He was a friend of the composer Carl Orff and wrote over three hundred works for chamber ensemble, orchestra, and chorus. As a true polymath, Bing also wrote several published works of fiction. In 1969, he moved to Washington University, where he continued his work on cardiac metabolism with Danforth as a partner.

Working with Bing, Danforth studied how the heart compensated for reductions in coronary flow and oxygenation. They found that the heart continues to beat so long as the membrane potential is maintained and action potentials can be generated. The study was extended to one in which the ability of the heart to utilize fatty acids was examined in both humans and dogs. In agreement with previous findings, they found that the heart utilized large amounts of free fatty acids, but a new finding was that esterified fatty acids accounted for more than half of the total fats extracted by the myocardium. They also made an extensive study of the metabolism of the failing heart in humans and dogs.

At Washington University, Danforth was appointed assistant and then associate professor of medicine, and during this time he received a postdoctoral fellowship in biochemistry to work in the Cori department. His first project was conducted with Ernst Helmreich and Carl Cori and concerned the effects of contraction and epinephrine on the phosphorylase activity of frog sartorius muscle. They found that the activation and deactivation of phosphorylase consequent to the addition and removal of epinephrine was much slower compared with the effects of contraction. The difference was a matter of minutes compared with seconds. A follow-up study with Helmreich looked at the conversion of phosphorylase b to phosphorylase a during contraction, and they concluded that changes in phosphorylase kinase activity were involved.

Research on Muscle Glycogen Metabolism

Danforth moved back to the Department of Medicine, where he was promoted to full professor and, in 1965, was appointed vice chancellor for medical affairs. He continued his research on the control of carbohydrate metabolism in muscle. Projects included measurements of glucose transport and phosphorylation in the dog heart. The uniqueness of the study was that an open chest approach was used in order to have the heart functioning as near to normal as possible. Another study carried out in collaboration with John B. Lyon Jr. of Emory University examined the effect of tetanic stimulation on glycogenolysis in muscles from two strains of mice, one of which lacked phosphorylase kinase activity and had very low levels of phosphorylase a. Glycogen was degraded in both strains, but in the normal strain, the maximal rate of glycogenolysis was greater and occurred earlier than in the mutant strain.

Danforth also studied the effects of electrical stimulation and epinephrine injection on the interconversion of glycogen synthase in muscles from anesthetized mice. Epinephrine caused a decrease in the I form of the enzyme and partially blocked the increase caused by tetanic stimulation. In mice that lacked phosphorylase kinase, glycogen synthase was almost entirely in the D form, and tetanic stimulation caused little increase in synthase I. An interesting relationship between glycogen levels and glycogen synthase activity was noted in these studies. This was that when tissue glycogen levels are low, the interconversion of glycogen synthase between its two forms favors glycogen synthesis, whereas high glycogen levels are associated with low synthesis. This relationship was also seen in the isolated rat diaphragm, where insulin was found to increase the activity of the I form of the enzyme. When the incorporation of ^{14}C-glucose into glycogen in the diaphragms was measured, they observed a correlation with activity of the I form of glycogen synthase.

Administrative Roles at Washington University

Because of his appointment as vice chancellor for medical affairs in 1965, Danforth had to taper off his research activities. He became chancellor of the entire university in 1971, and he held this position until 1995. Danforth was very popular with the students, who called him "Uncle Bill" or "Chan Dan," and he was frequently seen walking the campus and talking with the students. He was likewise popular with the faculty not only because of his accomplishments but also because of his relaxed and pleasant personality. During his tenure of twenty-four years, seventy new faculty chairs were established and the endowment increased significantly due to several major capital campaigns. Dozens of new buildings were erected, and the number of gift-supported scholarships tripled. After stepping down from the chancellorship, he became chairman of the board of trustees, and in 1999 they named him chancellor emeritus.

Danforth and his wife Elizabeth (Ibby) were responsible for numerous named scholarships, professorships, scholars programs, a butterfly garden, and a residence hall at Washington University. In 1971, he was elected to the Institute of Medicine of the National Academy of Sciences, and in 1982 he was elected to the American Academy of Arts and Sciences and also as a fellow of the American Association for the Advancement of Science.

Since retirement, he has been involved in the Donald Danforth Plant Science Center in St. Louis, where he is chairman of the board of directors and also of the Coalition of Plant and Life Sciences. In 2003 he was appointed by the Agriculture Secretary to chair the Research, Education, and Economics Task Force for the U.S. Department of Agriculture. The recommendations of the task force were focused on the need to strengthen food and agriculture research, especially fundamental research into basic plant and animal biology. Out of the recommendations, the National Institute of Food and Agriculture was established, with the function of evaluating and supervising competitive research grants at the Department of Agriculture. In addition to these activities, he continues to be involved in the Danforth Foundation

References

Ballard F. B., W. H. Danforth, S. Naegle, and R. B. Bing. 1960. "Myocardial Metabolism of Fatty Acids." *J. Clin. Invest.* 39: 717–23.
Danforth, W. H. 1965. "Glycogen Synthetase Activity in Skeletal Muscle. Interconversion of Two Forms and Control of Glycogen Synthesis." *J. Biol. Chem.* 240: 588–93.
Danforth, W. H., F. B. Ballard, K. Kako, J. D. Choudhury, and R. J. Bing. 1960. "Metabolism of the Heart in Failure." *Circulation* 212: 112–23.
Danforth, W. H., and R. J. Bing. 1958. "The Heart in Anoxia and Ischemia." *Brit. J. Anaesth.* 30: 456–65.
Danforth, W. H., and E. Helmreich. 1964. "Regulation of Glycolysis in Muscle. I. The Conversion of Phosphorylase *b* to Phosphoryase *a* in Frog Sartorius Muscle." *J. Biol. Chem.* 239: 3133–38.
Danforth, W. H., E. Helmreich, and C. F. Cori. 1962. "The Effect of Contraction and of Epinephrine on the Phosphorylase Activity of Frog Sartorius Muscle." *Proc. Natl. Acad. Sci. U.S.A.* 48: 1191–99.
Danforth, W. H., and J. B. Lyon Jr. 1964. "Glyogenolysis during Tetanic Contraction of Isolated Mouse Muscles in the Presence and Absence of Phosphorylase *a*. " *J. Biol. Chem.* 239: 4047–50.
Danforth, W. H., J. J. McKinsey, and J. T. Stewart. 1962. "Transport and Phosphorylation of Glucose in the Dog Heart." *J. Physiol.* 162: 367–84.
Grimes, W., 2010. "Obituary of Richard Bing." *New York Times*, November 14.
Johna, S. 2003. "Allen Oldfather Whipple: A Distinguishednd Historian." *Dig. Surg.* 20: 154–62.
Biographical Information, Donald Danforth Plant Science Center.
Biographical Information, Washington University.

29

The Influence of the Coris on Washington University and Carl Cori's Research at Boston

Retirement from Washington University

Carl Cori retired from the chairmanship of the Department of Biological Chemistry at Washington University in 1966 at the age of seventy and moved to Boston with his wife, Anne. Interestingly, he never returned to St. Louis. Philip Randle notes that, as a department head, Cori was primarily interested in research and had little interest in classroom teaching. He also did little committee work at the university and outside of the university. However, when he chose to utilize his political and administrative skills, they were judged outstanding. Both Carl and Gerty were true intellectuals with interests that ranged beyond science, into areas such as art, literature, and archeology. They were described by Philip Randle as having an insatiable appetite for knowledge and being collaborators in the pursuit of art and science. Chancellor Danforth said Carl left an indelible mark on the university by the example of the outstanding productivity of his group and the unexcelled quality of his trainees. Danforth noted that he and other administrators and professors often sought and heeded Carl's advice and counsel. His vision and breadth of interest made him influential in the selection of departmental heads. Danforth felt that the Coris were a major factor in elevating Washington University to the first ranks of academic institutions (Figure 26).

On retiring from Washington University, Carl displayed his erudition by paraphrasing some verses from the Roman poet Horace: *"Vixi scientia nuper idoneus / Et militavi non sine gloria"* ("Till now I have lived a life in service to Science and have fought my battles not without glory"). He added some personal philosophy and described episodes from the lives of Archimedes and Galileo in the context of extensive thoughts on the obligation of scientists to society. He commented that no scientist can wholly escape fulfilling some sort of obligation to society. Carl also noted that a scientist would have to reconcile himself to the fact that the self-correcting system under which science operates is not applicable to human affairs, and would step from his splendid isolation into the strong currents of human emotions. Carl described how he and Gerty enjoyed living in St. Louis, where they made many friends. He said that the university had given them splendid opportunities and that they could not be persuaded to move elsewhere.

Carl Cori was succeeded in the chair by Roy Vagelos, who had obtained his MD from the College of Physicians and Surgeons of Columbia University in 1954. After internship and residency at the Massachusetts General Hospital, Vagelos went to NIH to work with Earl Stadtman. He studied fatty acid synthesis, with a focus on acyl carrier protein. This protein is essential for fatty acid synthesis, because acetyl-CoA and malonyl-CoA are transferred to it before being condensed to β-ketoacyl-acyl carrier protein, which is further processed for fatty acid synthesis. At Washington University, Vagelos continued working on fatty acid metabolism and expanded his research to complex lipids and cholesterol. He left in 1975 to become senior vice president for research at Merck and was promoted to chief executive officer and chairman of the board there in 1984. Vagelos served in those positions until 1994. Under his leadership, the company flourished and expanded its philanthropy, for example, making one of their drugs free for the treatment of river blindness in Africa and Central America. He was also involved in the development of statins, which improve cardiovascular disease by inhibiting cholesterol synthesis. In 1972, he was elected to both the National Academy of Sciences and the American Academy of Arts and Sciences and has received many other awards.

Cori's Genetic Research at Boston

Cori described his experience living in Cambridge, Massachusetts, as happy and interesting. His marriage to his second wife Anne was a happy one, and they shared many diverse interests. Mildred Cohn states that Carl's wit and grace flourished during this last phase of his career. He was appointed visiting professor of biological chemistry at Harvard Medical School, with a laboratory at Massachusetts General Hospital. He remained active in research until his death in 1984. Cori's scientific interests took a new direction, namely the regulation of enzyme activity at the level of gene expression. This involved collaboration with Salomé Glüecksohn-Waelsch, a noted geneticist at Albert Einstein College of Medicine in New York.

The research was concerned with X-ray-induced mutations in the albino locus on chromosome 7 in mice. Because this locus contains genes that control the synthesis of liver enzymes, some of the animals displayed lethality due to suppression of glucose-6-phosphatase and a subsequent inability to maintain blood glucose levels. In the course of this work, efforts were made to purify glucose-6-phosphatase from liver microsomes, and Cori and his colleagues noted that there was a loss of activity when the enzyme was separated from phospholipids and that activity could be restored by adding back these lipids.

They also observed defects in the activity of other enzymes and a deficiency in plasma protein synthesis in the irradiated mice. Interestingly, they found the mutational effects on the proteins not to be exerted on their structural genes, which were found to be undamaged by the X-ray treatment, but were interpreted on the basis of cell hybridization studies as being due to loss of the genes that regulate their expression. The idea that regulatory genes were separate from structural genes—but that both were needed for enzyme expression—was a major breakthrough. This phenomenon was previously known in bacteria and yeast, but not mammals.

Cori continued to make seminal discoveries when he was in his eighties, and his mind continued to absorb and use new knowledge until the end of his life. He devoted some of his time to writing review articles and biographies of scientists he had known, for example, Otto Warburg and James Sumner, who had won the Nobel Prize in 1946 for the first crystallization of urease, demonstrating that it is a protein. Cori died at his home in Cambridge on October 20, 1984, at the age of eighty-seven.

The Coris received many honors during their careers. In addition to receiving the Nobel Prize in 1947, they were elected to the National Academy of Sciences. Most of the awards were given jointly, but because of Gerty's premature demise, Carl received several alone. These included many honorary degrees from prestigious universities and election as a Foreign Member of the Royal Society and of the Danish Royal Society. He was also a member of the American Philosophical Society and received the Willard Gibbs Medal of the American Chemical Society. He was president of the American Society of Biological Chemistry and president of the Fourth International Congress of Biochemistry at Vienna in 1958. He was a member of the Scientific Committee of the Helen Hay Whitney Foundation, and this reflected his interest in the careers of young scientists. On September 21, 2004, both Carl and Gerty were designated an American Chemical Society Landmark in recognition of their work in elucidating carbohydrate metabolism.

References

Cohn, M. 1992. "Carl Ferdinand Cori 1896–1984." *Biogr. Mem. Natl. Acad. Sci. U.S.A.* 77–109.
Cori, C. F. 1969. "The Call of Science." *Ann. Rev. Biochem.* 38: 1–21.
Cori, C. F., S. Glüecksohn-Waelsch, H. P. Klinger, L. Pick, S. L. Schlagman, L. S. Teicher, and H. F. Wang-Chang. 1981. "Complementation of Gene Deletions by Cell Hybridization." *Proc. Natl. Acad. Sci. U.S.A.* 78: 479–83.
Cori, C. F., S. Gluecksohn-Waelsch, P. A. Shaw, and C. Robinson. 1983. "Correction of a Genetically Caused Enzyme Defect by Somatic Cell Hybridization." *Proc. Natl. Acad. Sci. U.S.A.* 80: 6611–14.
Erickson, R. P., S. Glueksohn-Waelsch, and C. F. Cori. 1968. "Glucose-6-Phosphatase Deficiency by Radiation-Induced Alleles at the Albino Locus in the Mouse." *Proc. Natl. Acad. Sci. U.S.A.* 59: 437–44.
Garland, R. C., and C. F. Cori. 1972. "Separation of Phospholipids from Glucose-6-Phosphatase by Gel Chromatography." *Biochemistry* 11: 4712–18.

Kresge, N., R. D. Simoni, and R. L. Hill. 2005. "The Role of Acyl Carrier Protein in Fatty Acid Synthesis: The Work of P. Roy Vagelos." *J. Biol. Chem.* 280: e32.

Majerus, P. W., A. W. Alberts, and P. R. Vagelos. 1964. "The Acyl Carrier Protein of Fatty Acid Synthesis: Purification, Physical Properties, and Substrate Binding site." *Natl. Acad. Sci. U.S.A.* 51: 1231–38.

Randle, P. J. 1986. "Carl Ferdinand Cori 5 December 1896–20 October 1984." *Biogr. Mems. Fell. R. Soc.* 67–95.

30

The Heritage of the Coris

Characteristics of the Coris and their Laboratory

The influence of the Coris on biochemistry extended beyond their department to encompass other institutions in the United States and the world. It was due not just to their own achievements, but to the accomplishments of their students through the excellent training they received. Although it cannot be claimed that the discoveries of the Nobel Laureates and other distinguished scientists who worked in their laboratory are entirely attributable to the Coris, the rigorous training in the scientific method that they received played a role. As noted in the previous chapters, the majority of the Coris' trainees became major figures in biochemistry, and many became departmental chairmen in major medical schools. Their influence on biochemistry in America was immense and extended overseas to the many countries from which their trainees came and returned. Their impact was felt in innumerable ways, and their findings and ideas altered thinking in the discipline. Interestingly, despite the fact that Carl and Gerty were both MDs, their research was entirely basic, apart from their work on glycogen storage diseases. Like Arthur Kornberg, they were convinced that any advances in medicine would likely come from basic untargeted research. Although many of their trainees took positions in clinical departments, they engaged in both basic and clinical research, bringing to both types of research the approach and rigor that were inculcated in them by the Coris.

It is interesting to reflect on how two Austrian émigrés born at the end of the nineteenth century, with no formal training in biochemistry and forced, because of the rise of Nazism, to take positions in a second-rate research institute in Buffalo, nevertheless rose to achieve preeminent positions in American science. One factor was their academic heritage. Carl came from a family that was distinguished by the presence of many professors, and Gerty came from a cultured background. Another point was that they achieved much because they worked as a great team, despite considerable differences in their personalities. Carl, with his calm and analytic demeanor, was the principal intellectual force, and Gerty, with her excitable but intuitive nature, was the gifted experimentalist. Carl commented that their efforts were complementary and that neither could have gone so far without the other. However, despite the synergy between Carl and Gerty, several attempts were made to break them up as a team early in their careers.

Another reason for their success was their command of the literature, which was both broad and deep and which was facilitated by their fluency in several languages and the close proximity of their library. At the time they were at Washington University, all departments had their own libraries. In the case of the Coris, the library was right next to the laboratory, which was next to Carl's office. This indicated the importance they placed on keeping up with the literature. The library was where the seminars were presented, usually over lunch. Trainees were expected to present their findings to the entire group, and departmental members were required to attend; apologies were expected if absence was unavoidable. Visiting scientists would always be invited to present their findings or comment on the work of departmental members, and Gerty would bring lunch for them. The discussion could involve other new scientific findings or even scientific gossip. Seminar speakers used either a blackboard or glass projection slides. On all of these occasions, Carl's formidable intellect and memory would be on display, and Gerty would signify her positive or negative impressions of the talks by the appropriate facial expressions. This was all very intimidating to the junior members of the laboratory. Despite Carl's towering presence, he was tolerant of the failings of others and endeavored to promote an atmosphere of collaboration and mutual support within the laboratory. He always acknowledged the contributions of his collaborators and was scrupulously fair in any discussions.

As noted, Carl and Gerty had an insatiable appetite for knowledge. Their reading was by no means restricted to science and reflected their wide interests. Carl had an interest in music through his maternal grandfather and learned to play the cello as a young man until his cello was broken. Both Gerty and Carl were theatergoers and concertgoers and were devoted to their garden. From their early careers, mountain climbing and gardening were some of their pursuits and represented their principal physical activities. In contrast, they had no interest in spectator sports. They recognized the importance of social interactions in the department, and they were by no means antisocial and entertained in their home often. Periodically, they would hold an open house to which the whole department would be invited.

Approach to Research

Prior to their time in St. Louis, Carl and Gerty worked together to make some of their major discoveries with almost no collaborators and in a less than stellar environment. During this time, it was evident that their approach to research was most rigorous, and their high standards would continue and be communicated to their students. Through their great command of the literature, they made certain that the questions asked were truly novel. They also gave great attention to experimental details and design, making sure to exclude any artifacts and validating any analytic procedures. Sufficient animals and preparations were used to insure that any results were significant and reproducible. A critical point in their studies was that the results were not just descriptive, but were developed into hypotheses. Finally, any emerging hypotheses were subjected to the most careful scrutiny, and, whenever possible, the results were subjected to quantitative analysis. The utilization of quantitation meant that some hypotheses had to be discarded and replaced by others where the numbers made sense. They were great exponents of the scientific method and abhorred loose thinking by their associates. They discouraged them from generating information that was already known or working on projects that they considered uninteresting or nonsignificant. Adherence to these principles and the consequent impact and reliability of their findings greatly enhanced their reputations and that of their laboratory.

Through their command of the German language, the Coris were familiar with the biochemical research emerging from some of the great European laboratories prior to World War II, in particular those of Otto Warburg and others at Berlin-Dahlem, which focused on the reactions and enzymes involved in the metabolism of glucose. The work of Warburg drew them to consider the hormonal regulation of glucose and lactic acid metabolism, and led to the formulation of the renowned Cori cycle. Their investigations into carbohydrate metabolism utilizing intact animals were the most novel, extensive, and thorough of any carried out in the United States at that time. They included studies of the absorption of glucose from the intestine and what happened to this sugar in the body, namely, how much was oxidized and how much was converted to glycogen. They also studied the effects of epinephrine and insulin on these pathways, although their findings with insulin were complicated by the unknown presence of glucagon in their insulin preparations. However, their studies of the metabolic effects of epinephrine were of great physiological significance and had implications for the metabolic changes during exercise. Surprisingly, despite the excellence of their research, their conclusions were not universally accepted, because they ran counter to the views of some of the major researchers in the area.

An important turning point in their careers, which coincided with their moving to St. Louis in 1931, was when they recognized that working with whole animals presented many experimental difficulties that frequently made it difficult to devise clear-cut experiments and draw unequivocal conclusions. They realized that they needed to obviate these problems by switching to isolated tissues and ultimately to pure enzymes. This was a rather courageous step since they did not have experience in these systems, and the new department did not have the equipment and facilities for this type of research. Many investigators would have resisted changing to new research utilizing new procedures, but not the Coris. As a result of their switch to studies of purified enzymes, they accomplished the research for which they were awarded their Nobel Prizes.

Because the vast amount of research accomplished at Washington University has already been presented in the preceding chapters, the focus here will be on some of the major accomplishments of their laboratory. Their initial work in St. Louis centered on phosphorylase and resulted in the discovery with Sidney Colowick that the product of its action on glycogen was a novel hexose phosphate, namely glucose-1-phosphate. They also found that the enzyme could synthesize glycogen in vitro and concluded that this was the way this macromolecule was synthesized in vivo. However, the subsequent work of Luis Leloir showed that another enzyme, glycogen synthase, was responsible. Nevertheless, the demonstration that phosphorylase could synthesize glycogen represented the first example of the synthesis of a macromolecule in a cell-free system. The demonstration of the role of phosphorylase in the metabolism of glycogen won the Coris the Nobel Prize in 1947.

Leloir's research in the Cori laboratory involved studies of citrate synthase, a key enzyme in the citric acid cycle. His transformative research on the synthesis of glycogen and other complex carbohydrates, which won him the Nobel Prize in 1970, was done when he returned to Argentina. During his peripatetic journey through the United States, Herman Kalckar stopped in St. Louis and worked with Sidney Colowick and described his work with Colowick as some of his happiest months in research. They studied yeast hexokinase and found a contaminant that turned out to be a novel enzyme, namely, adenylate kinase.

With Arda Green, the Coris proceeded to purify and crystallize phosphorylase and characterize it in great detail. A transformative discovery was that it existed in two forms that differed in activity and that could be interconverted. The idea that an enzyme could exist in two interconvertible forms was a truly novel finding. Carl attributed the higher activity of phosphorylase *a* to the presence of a prosthetic group, but later work by Earl Sutherland and Edwin Krebs showed it was due to phosphorylation. Another significant finding was that the activity of the less active form could be stimulated by AMP. This was, in fact, the first demonstration of the allosteric regulation of an enzyme, although this phenomenon was not characterized until many years later. Another important enzyme that they purified and studied in detail with Colowick and Victor Najjar was phosphoglucomutase, whose unique mechanism of action was studied with Earl Sutherland, Mildred Cohn, and Theo Posternak. They also studied glyceraldehyde-3-phosphate dehydrogenase, which is a complex enzyme involved in the generation of ATP in the glycolytic pathway. It was crystallized and characterized in association with Milton Slein, Sidney Velick, and Jane Park. Mildred Cohn went to the Cori laboratory in 1946. She initially decided to work independently of Carl, but after discussions with him, resolved to utilize ^{18}O to analyze the reaction mechanisms for phosphorylase, phosphatases, and glyceraldehyde-3-phosphate dehydrogenase. The results revealed striking differences in their catalytic mechanisms. She later utilized NMR to deduce structural features of the binding sites of kinases.

In studies of the effect of insulin on glycogen breakdown in the liver, Carl, together with Sutherland and Christian de Duve, made the paradigm-shifting observation that this was due to a contaminant (glucagon) that activated phosphorylase. Interestingly, Carl was initially reluctant to accede to a request from de Duve to work in his laboratory since he did not agree with his view of how insulin acted, but he later relented. Sutherland and de Duve extended their findings with glucagon to show that epinephrine had a similar effect and that the activation of phosphorylase was due to an increase in the active form. These observations led to discoveries of great importance in the laboratories of Sutherland and Edwin Krebs relating to the fundamental role of cyclic AMP in the actions of many hormones, neurotransmitters, and drugs and the importance of phosphorylation in regulating the activity of many enzymes. Sutherland won the Nobel Prize in 1971 and Krebs in 1992. Somewhat surprisingly, Krebs did not make any significant discoveries in the Cori laboratory. He examined the effects of protamine and AMP on phosphorylase, but Carl was reluctant to publish the findings, because he considered them to be phenomenological. As described above, it was not recognized that these represented a way of controlling enzyme activity through allosteric effects. On his return to Belgium, de Duve won the Nobel Prize in 1974 for discovering lysosomes, a new type of intracellular organelle.

In addition to their studies of phosphorylase activation, the Coris studied how the glycogen molecule itself was degraded, and this led to the discovery that phosphorylase alone could not degrade the molecule completely and that a second enzyme (amylo 1,6-glucosidase), which dealt with the branch points, was required. Gerty then proceeded to use these enzymes to determine the branched structure of glycogen and to use the knowledge of these and other enzymes to analyze the basis for several glycogen storage diseases. This work was carried out in collaboration with Joseph Larner and David and Barbara Brown née Illingworth. During this phase of her career, Gerty exhibited great determination since she did much of this work when she was exhausted by the anemia that led to her death in 1957.

In the mid-1960s, Carl turned his attention to the effects of insulin and other hormones on glucose metabolism in muscle. His earlier foray into this area involved the demonstration that adrenocortical and pituitary factors had an inhibitory effect on hexokinase in vitro, which could be reversed with insulin. However, to the great distress of Carl and Sidney Colowick, these effects could not be reproduced, and it appeared that the findings had been falsified by one of their collaborators. The study of the effects of hormones on muscle metabolism was transferred to another experimental system, namely, the isolated rat diaphragm, and many young investigators, including Charles Park, David Kipnis, Ernst Helmreich, William Daughaday, William Danforth, David Brown, Mike Krahl, and Marvin Cornblath, were involved at various stages. One of the main conclusions was that insulin acted by stimulating the transport of glucose into the muscle cells and not its phosphorylation. However, Carl initially resisted this conclusion and felt that any increase in uptake was due to some intracellular event. Later experiments showed that the inhibitory effects of glucocorticoids and growth hormone on the stimulation of glucose uptake by insulin were due to effects on its phosphorylation. Most of the investigators had subsequent spectacular careers in academic medicine. For example, Park did fundamental work on the mechanisms of action insulin and the control of carbohydrate metabolism; Helmreich become a leading figure in German biochemistry; Kipnis became an authority on diabetes; Daughaday became renowned for his work on growth hormone; Danforth rose to become the chancellor of Washington University; and Cornblath became recognized for his work on neonatal metabolism.

Several other scientists worked in the Cori laboratory and then proceeded to have interesting careers. One was Robert Crane, who discovered glucose-Na^+ cotransport in the intestine. He was also notable for devising oral rehydration (the oral administration of salt and glucose) as an effective treatment of diarrhea due to cholera. This simple procedure alleviated a condition suffered by millions in Africa, India, and other Eastern countries. Alberto Sols worked with Crane studying the kinetics of hexokinase and became a major figure in Spanish science. Luis Glaser, who escaped from Nazism in Austria, made major contributions to the biology of complex carbohydrates. He trained some important figures in biochemistry and became chairman of biological chemistry at Washington University before moving to the University of Miami as executive vice president and provost. Carl Frieden was another recruit of Carl Cori who was a star in the field of enzyme kinetics and served as chairman of the department from 2000 to 2005. As of this writing, he is still doing research at over eighty years of age.

One of the most famous biochemists to work in the Cori laboratory was Arthur Kornberg, who won the Nobel Prize in 1959. Kornberg had fallen in love with enzymes during his time with Severo Ochoa and went to the Cori laboratory because he considered it the mecca of enzymology. His project was to find out why liver particles produced pyrophosphate and found it to be produced by the cleavage of NAD. The enzyme involved seemed rather boring, but Kornberg remembered his famous dictum that there is never a dull enzyme. He was right, because the enzyme led to another that synthesized two major coenzymes, and its mechanism of nucleotidyl transfer from a nucleoside triphosphate was found to be utilized repeatedly in the synthesis of proteins, lipids, carbohydrates, and nucleic acids.

After spending time at NIH, Kornberg returned to Washington University as chairman of microbiology and began a career that resulted in the most significant accomplishments of any of the trainees of the Coris. This was the discovery and purification of DNA polymerase and the demonstration that it could proofread and edit its product and, most important, that it was capable of generating a fully functional phage. This latter accomplishment was only accomplished after almost ten years and was labeled as the generation of "life in a test tube." As in the case of the Coris, Kornberg's work with DNA polymerase was initially assailed by some in the scientific community, and his seminal papers were rejected for publication at first. However, his work soon became recognized as revolutionary and today has formed the basis of genome sequencing, the production of medicinal agents through recombinant technology, the accomplishment of mutagenesis, and the production of transgenic animals. Kornberg's work also led to the development of the polymerase chain reaction, which is used to analyze a large variety of physiological, pathological, and forensic conditions.

It could be asked, what did the Coris contribute to Kornberg's accomplishments? The answer is their intense focus on enzymes. This involved their purification, characterization, and, when possible, their crystallization. This was important because all of Kornberg's critical discoveries depended on the availability of pure enzymes. Another point was the Coris' emphasis on the mechanisms underlying the actions of enzymes. Finally, Kornberg's studies in the Cori laboratory on the origin of cellular pyrophosphate led to the discovery of enzymes that synthesized coenzymes and purine nucleotides, the latter being required for nucleic acid synthesis.

The Cori laboratory was known for the astonishing number of Nobel Laureates who worked there, but the list of members elected to the National Academy of Sciences is also impressive. These scientists have, in turn, trained a cadre of remarkable scientists. These include two Nobel Laureates, Alfred Gilman and Ferid Murad, and twelve members of the National Academy. Thus, the scientific ethos of the Coris and the number of researchers spawned from their scientific crucible continue to this day.

Index

Abumrad, Nada, 136
Acetoacetate, 33
Acetyl CoA, 33, 189, 211
Acetyl CoA carboxylase, 37
Acetylation, 198
Acromegaly, 167
Acyl carrier protein, 211
Adenosine monophosphate (AMP) 13, 39, 70, 71, 84, 190, 217
Adenosine 3',5'-monophosphate (cyclic AMP) 57–59, 62, 86, 132, 135, 153–156, 203
Adenosine diphosphate (ADP) 17, 97, 173
Adenosine diphosphate glucose (ADPG) 51
Adenosine triphosphate (ATP) 17, 34, 97, 142, 143
Adenylate cyclase 58, 59, 155, 192, 193
Adenylate kinase, 17, 25, 217
Adrenal gland, 17, 42, 58, 123, 124, 130, 165, 167, 218
Adrenergic receptor, 58, 62, 135, 136, 165, 167, 218
Adrenocorticotrophic hormone (ACTH), 58, 61
Alanine, 133, 135, 204
Alberty, Robert, 196, 197
Aldolase, 31, 42, 73
Allosterism 40, 84, 173, 190, 197
Amino acid, 48, 94, 95, 133, 203
Amino acid sequence, 73, 153
Amylo-1,6-glucosidase, 145, 146, 150, 163, 218
Amylo-1,4→1,6-transglucosidase, 163
Amylopectin, 150, 162
Anemia, 70, 162
Angiotensin, 42, 47
Angiotensinogen, 42, 47
Anoxia, 130, 131
Antibody, 79
Antigen, 79
Austrian Army, 3, 4
Austro-Hungarian Empire, 1

Ball, Eric, 75
Banting, Fred, 8
Barnes Hospital, 54, 83, 167, 185
Bates, Dave, 198
Beavo, Joe, 132
Beckman DU spectrophotometer, 113
Berg, Paul, 26
Berlin-Dahlem, 7, 9, 232, 216
Bernard, Claude, 7
Berthet, Jacques, 56, 107
Best, Charles, 8, 106
Bilirubin, 50, 112
Bing, Richard, 206, 207
Biotin, 37
Bohr, Niels, 21
Bohr, Christian, 21
Bond, Judith, 143
Bornstein, Joe, 123, 125, 129
Bouckaert, Joseph, 100, 101
Boyer, Paul, 96
Broadus, Arthur, 61
Brown, Barbara née Illingworth, 71, 145, 161–163, 218
Brown, David, 123, 161–163, 181, 218
Butcher Reginald W. (Bill), 59, 61, 63, 132

Cahill, George, Jr., 203
California Institute of Technology (CalTech), 24 117
Calmodulin, 184
Cambridge University, 22, 32
Campomar, 48–50, 153
Cannon, Walter, 10
Cantoni, Guilio, 36
Carbohydrate metabolism, 9, 124, 133–135, 162, 202, 216, 219
Carlsberg Laboratory, 22, 24, 27

Catalytic subunit, 27, 135
Cell wall, 51, 186
Cellulose, 51, 161
Chain, Ernst, 31
Chance, Britton, 97, 104, 140, 141
Cherrington, Alan, 134, 135
Chitin, 50, 161, 181
Citric acid cycle, 22, 32, 113
Clarke, Hans, 36
Claude, Albert, 106–109
Cleveland Clinic, 42
CoenzymeA (CoA), 33
Cohn, Mildred, 36, 55, 70, 91–99, 142, 173, 217
Cohn, Edwin J, 41
Colman, Roberta, 198, 199
Colowick, Sidney, 16–19, 24, 83, 121, 123, 217, 218
Complex carbohydrates, 51.181, 186
Cori, Anne, 146, 211
Cori, Carl, 1–16, 24, 31, 39, 40, 53–5, 67–71, 73, 78, 84–86, 104, 105, 114, 116, 123, 128, 146, 150, 172–174, 190, 193, 201, 207, 214–220
Cori, Gerty, 3–16, 24, 31, 39, 67–71, 83, 84, 114, 116, 129, 139, 145–147, 150, 214–220
Cori, Tom, 68, 69
Cori cycle, 10
Cori ester, 14
Cornblath, Marvin, 123, 125, 161, 218
Cortisol, 130, 219
Coumadin, 196
Crane, Robert, 171–176, 219
Creatine phosphate (Phosphocreatine), 22, 29, 30, 142, 143
Creatine phosphokinase, 183
Crigler-Najjar syndrome, 79
Cyclic AMP see Adenosine 3',5'-monophosphate
Cyclic AMP-dependent protein kinase (PKA), 41, 57, 86, 153–155
Cyclic GMP see Guanosine 3',5'-monophosphate
Cyclic GMP-dependent protein kinase (PKG), 135
Cytochromes, 32, 75

Dale, Sir Henry, 6, 30
Dali, Salvador, 29, 38
Danforth, William, 123, 190, 206–210, 218
Daughaday, William, 123, 151, 161, 165–169, 202, 218
Delbrück, Max, 24, 27
De Duve, Christian, 54, 100–110, 208
Deoxynucleoside triphosphates, 115–117, 219
2-Deoxyglucose, 124, 201, 202
Deoxyribonucleic acid (DNA), 115–116
Dephosphorylation, 56
Depression, Great, 16, 82, 89, 116
Deuterium, 91, 92, 94
Diabetes, 75, 124, 128, 130, 134, 167, 168, 185
Diaphragm, 123, 128, 130, 201, 208, 218
Diarrhea, 174

Dixon, Malcolm, 46
DNA polymerase, 116–119, 219
DNA replication, 118–120
Dolichol, 52
Donald Danforth Plant Science Center, 208
Du Vigneaud, Vincent, 93–95
Dwarfism, 168

Edsall, John, 42, 117
Edson, Norman, 46
Edwards, Jonathan, 127
Eidgenössische Technische Hochschule (ETH), 140, 191
Eigen, Manfred, 76, 191
Electron transfer chain, 32
Electron paramagnetic resonance (EPR), 95, 142
Eli Lilly Company, 103, 106
Embden, Gustav, 30
Endoplasmic reticulum, 51, 82
Enzyme Club, 35
Enzyme purification, 40, 113, 116, 220
Enzyme Section (NIH), 115
Epinephrine, 7–11, 13, 55–60, 62, 101, 123, 133, 135, 153, 154, 201, 207, 216
Erectile dysfunction, 136
Erythrocyte, 142,
Escherichia coli, 118, 119
Evans, Herbert, 41
Exercise, 143
Exton, John, 61, 133

Fatty acid oxidation, 76. 203
Fight or flight, 10
Fischer, Edmond, 56, 85, 87, 192
Fischer, Emil, 7
Flavin adenine dinucleotide (FAD), 7, 115
Fatty acids, 131, 136
Francis, Sharron, 135, 136, 142
Frieden, Carl, 75, 196–199
Friedkin, Morris, 25, 32, 116
Fructose metabolism, 31
Fructose-1,6-bisphosphate, 31
Fructose-1,6-bisphosphatase, 134
Fructose-2,6-bisphosphate, 134
Fructose-2,6-bisphosphatase, 134

Galactose, 27, 49, 173
Galactose metabolism, 26
Galactosemia, 27, 47
Galactose-1-phosphate, 49
Gale, Ernest, 79
Garber, Alan, 204
Garbers, David, 132
Gasser, Herbert, 13
Genetic code, 34
Gastric mucosa, 53, 54
German University of Prague, 1, 3

Index

Gilman, Alfred, 57, 132, 154–156, 193, 220
Glaser, Luis, 161, 181–187, 190, 219
Glucagon, 55–58, 60, 61, 106, 133, 135, 202, 218
Gluconeogenesis, 31, 125, 133, 135, 179, 203
Glucosaminoglycan, 50
Glucose, 7–10, 12, 133, 150, 201
Glucose sodium cotransport, 173
Glucose transport, 123, 124, 129–131, 207
Glucose phosphorylation, 17, 124, 130, 201, 207
Glucose-1-phosphate, 14, 16, 17, 40, 49, 50, 55, 70, 96, 150, 152
Glucose-1,6-bisphosphate, 49, 70
Glucose-6-phosphate, 14, 50, 55, 70, 106, 124, 130, 131, 152, 173, 177
Glucose-6-phosphatase, 55, 148, 211
Glucose-6-phosphate dehydrogenase, 161, 181
Glüecksohn-Waelsch, Salomé, 211
Glutamate dehydrogenase, 197, 198
Glyceraldehyde-3-phosphate dehydrogenase, 70, 73, 76, 84, 85, 139, 141, 142, 172, 217
Glycobiology, 50
Glycogen, 7, 9, 11, 13, 17, 23, 40, 50, 54, 70, 124, 145, 162, 207, 217, 218
Glycogen synthase, 500, 87, 152–154, 158, 208, 217
Glycogen storage diseases, 108, 145, 146, 151, 163, 218
Glycolysis, 21, 23, 131, 179
Golgi, 182
Gordon, Jeffrey, 184–185
G proteins, 57, 155, 156, 192
Green Arda, 41–43, 83
Green, David, 46, 48
Grisolia, Santiago, 32
Growth factors, 186
Growth hormone, 125, 130, 167, 219
Grunberg-Manago, Marianne, 34, 98, 116
Guanosine 3',5'-monophosphate (Cyclic GMP), 60–62, 132, 135, 136, 157, 203
Guanosine diphosphate-mannose (GDP-mannose), 50
Guanosine diphosphate-glucose (GDP-glucose), 50
Guanosine triphosphate (GTP), 155, 192, 193
Guanylate cyclase, 132, 157

HAND proteins, 184
Hardman, Joel, 60, 62–65, 132
Heard, Alexander, 19
Heart, 130, 131, 184, 207
Heimberg, Murray, 74, 75
Helmreich, Ernst, 76, 123, 189–194, 201, 218, 219
Heppel, Leon, 18, 57
Hers, Henri-Géry, 31, 107, 108, 134
Hexokinase, 17, 25, 70, 79, 123, 125, 130, 172, 173, 177, 217
Hill, Archibald V., 22, 31, 100

Holt, Emmett, 78
Hopkins, Sir Frederick G., 9, 22, 46, 151
Houssay, Bernardo, 69, 71, 128, 147
Huxley, Aldous, 103
Hyaluronic acid, 50 181
Hyperglycemic-glycogenolytic factor, 54
Hypertension, 42, 47, 136
Hypoglycemia, 11, 101, 202

Ibsen, Henrik, 21
Illingworth, Barbara, see Barbara Brown
Inositol, 151, 158, 167
Insulin, 8, 11, 12, 17, 54, 61, 101–104, 123, 128, 129, 133, 135, 152, 158, 162, 202, 203, 216, 218, 219
Insulin synthesis, 203, 204
Insulin-like growth factors, 168, 169
International Institute of Cellular and Molecular Pathology, 110
Isoprene, 31
Isselbacher, Kurt, 27

Jefferson, Leonard, 61, 132
Jewish scientists/refugees, 7, 21, 30, 181, 186
Johns Hopkins University, 18, 19, 27, 41–43, 73, 78, 79, 127
Johnson Foundation, 97, 98
Johnson, Louise, 192
Johnson, Roger, 61

Kafka, Franz, 5
Kaiser Wilhelm Foundation/Institute, 7, 23, 30
Kalckar, Herman, 17, 21–28, 48, 64, 217
Kamen, Martin, 166
Kaplan, Nathan, 16, 18
Keilin, David, 32
Kennedy, Eugene, 27, 32, 86
Ketone bodies, 33, 75, 125
Khorana, H. Gobind, 35
Kipnis, David, 123, 169, 201–204, 218, 219
Klenow, Hans, 25
Kono Tetsuro, 131
Kornberg, Arthur, 18, 26, 32, 37, 112–121, 181, 214, 219
Kornberg, Sir Hans, 71
Kornfeld, Stuart, 162, 182
Kornfeld, Rosalind née Hauk, 161, 162
Koshland, Daniel, 141
Krahl, Mike, 123, 125, 128, 218
Krebs, Edwin, 56, 82–89, 192, 217
Krebs, Sir Hans, 22, 36, 71, 96, 135
Krogh, August, 21
Kunsthistorisches Museum, 4

Lactic acid, 7–11, 13, 133
Lactose, 49
Lane, M. Daniel, 37

Larner, Joseph, 71, 146, 149–158, 162, 163, 218
Lawrence, Ernest, 92
Lehman, Robert, 116
Lehninger, Albert, 32
Leloir, Luis, 26, 39, 45–52, 70
Levine, Rachmiel, 130, 201
Levitzi, Alex, 193
Liddle, Grant, 132
Linderstrom-Lang, Kaj, 24 , 26
Link, Paul, 196
Lipase, 87
Lipkin, David, 57
Lipmann, Fritz, 22, 106, 141, 172
Lipolysis, 58
Lippich, Ferdinand, 1
Loewi, Otto, 6
Lohmann, Karl, 22, 23
Louvain/Leuven, 100, 102, 103, 109
Lowry, Oliver, 25, 121, 151
Luciferin, luciferase, 18
Lundsgaard, Eijnar, 21, 25, 130, 173
Lynen, Feodor, 33, 131, 189
Lysosomes, 108, 163, 182

Macloed, John J.R., 8
Madsen, Neil, 70
Mallette, Larry, 133
Malonyl-CoA, 37, 211
Mannose-6-phosphate, 182
Marine Biological Stations, 1, 3, 31, 141, 172,
Martius, Carl, 141, 191
Massachusetts General Hospital (MGH), 18, 27, 141, 165, 172, 211
Mass spectrometer, 92, 96
McCollum-Pratt Institute, 18, 27, 42
McElroy, William, 18, 27, 42
Mehler, Alan, 32, 33, 113
Meng, Ray, 75, 129
Merck, 211
Methods in Enzymology, 18
Meyer, Hans Horst, 6
Meyerhof, Otto, 9, 23, 30
Michaelis-Menten equation, 40, 97, 177, 197
Microbiota, gut, 185
Millikan, Glenn, 97
Mitchell, Peter, 174
Mitochondria, 48, 96
Mitogen-activated protein kinase (MAPK), 88
Monod, Jacques, 40, 190, 197
Morgan, Howard, 131, 133
Muñoz, Juan, 46
Murad, Ferid, 60, 156–158, 220
Muscle, 30, 123, 142, 143, 201, 207, 208
Muscular dystrophy, 142, 143
MyoD, 184

Myristoylation, 185

Na+-K+-ATPase (Na+ pump), 174
Najjar, Victor, 39, 70, 78–81, 140, 217
National Institutes of Health (NIH), 18, 112, 115, 149, 194
Naturhistorisches Museum, 4
Nazis/Nazism, 23, 30, 36, 92, 187, 214
Neurath, Hans, 85
New York University (NYU), 31–37
Nicotinamide adenine dinucleotide (NAD), 7, 18, 74, 76, 115, 141, 142, 197–199
Nicotinamide adenine dinucleotide phosphate (NADP), 18, 197–199
Nicotinamide mononucleotide, 115
Nicotinic acetylcholine receptors, 183
Nirenberg, Marshall, 35
Nitric oxide, 60, 156, 157
Non-suppressible insulin-like activity (NSILA), 168
Nuclear fallout, 27
Nuclear Magnetic Resonance (NMR), 95, 96, 98, 142, 199, 217

Obesity, 168, 185
Ochoa, Severo, 27–38, 113, 116, 140, 178
Olson, Eric, 183, 184
Oral rehydration, 174
Oxford University, 23, 31, 192
Oxidative phosphorylation, 24, 32, 96, 141, 172

Page, Irvine, 42, 47
Pancreas, 54, 106, 203
Parasympathetic nervous system, 62
Parathyroid hormone, 62
Park, Charles (Rollo), 61, 74, 78, 123, 127–137, 139, 140, 161, 202, 218, 219
Park, Edwards, 78, 127
Park, Jane née Harting (Janey), 74, 78, 129, 139–143
Parnas, Jacob, 13, 39
Pauling, Linus, 24, 41
Penicillin, 27, 31
Pentose, 124
Permutt, Alan, 203, 204
Peronista government, 48
Peroxisome, 109
Peter Bent Brigham Hospital, 128
Peters, Sir Rudolph, 31
Pfeuffer, Thomas, 192
Phage, 117–119
Phagocytosis, 80
Phosphodiesterase, 59, 61, 136
6-Phosphofructo-1-kinase, 131, 199
6-Phosphofructo-2-kinase, 134
Phosphoglucomutase, 17, 39, 49, 55, 70, 79, 80, 96, 152, 217

Phosphoglycerate mutase, 55, 70
5-Phosphoribosyl-1-pyrophosphate (PRPP), 115
Phosphorylase, 13, 39, 40, 55, 56, 70, 83–86, 96,
    105, 145, 150, 162, 186, 190–192, 207, 217, 218
Phosphorylation cascade, 88
Phosphorylase kinase, 41, 56, 86, 153, 207
Phosphorylase phosphatase 41, 86
Pilkis, Simon, 134
Piuitary gland, 17, 123, 124, 128, 129, 162, 218
Placental lactogen, 168
Polymerase chain reaction (PCR), 120
Polynucleotide phosphorylase, 34
Polypeptides, 51
Polysaccharides, 51, 150
Post, Robert, 129
Posternak, Theo, 55, 217
Prague, 51
Pressman, Burton, 107
Primakoff, Henry, 90
Primosome, 119
Propionyl CoA, 34, 37
Protein kinase, 96, 135, 158, 217
Protein phosphatase, 86, 96
Protein tyrosine kinase (Tyrosine kinase), 88
Protein phosphorylation, 86
Pullman, Maynard, 18
Pyridine nucleotide see Nicotinamide adenine
    dinucleotide
Pygmies, 168
Pyridoxal-5-phosphate, 162, 191, 192
Pyrophosphate, 51, 58, 115, 116, 152, 220

Racker, Efraim, 33
Rall, Theodore (Ted), 57
Ramón y Cajal, Santiago, 29
Randle, Sir Philip, 131, 220
Ratner, Sarah, 36
Rawlinson, Sir Henry, 127
Recombinant technology, 120
Regulatory subunit, 87, 135
Regen, David, 131
Renin, 42, 47
Replication, 118, 119
Replisome, 119
Ribonucleic acid (RNA), 34, 35, 98,
    119, 184
Rittenberg, David, 36, 92
Robison, G. Alan (Al), 59, 61, 63
Roche Institute of Molecular Biology, 35
Rose, Irwin, 150
Rose, W.C., 94
Rosell-Perez, Maria, 152, 153
Roswell Park Memorial Institute, 6
Rous sarcoma virus, 88

Salmon, William, 167
San Pietro, Anthony, 18

Schoenheimer, Rudolf, 36, 92
Second messenger hypothesis, 60
Schultz, Günter, 60
Selinger, Svi, 155
Serotonin, 42
Shaffer, Philip, 12
Sigma Chemical Company, 69
Singer, Maxine, 57
Sinsheimer, Robert, 57
Sneyd, John G.T. (Sam), 61, 132
Sols, Alberto, 172, 173, 176–179, 219
Somatomedin, 167, 169
Soskin, Samuel, 103
Starch, 50, 51
Starvation, 202
State Institution for the Study of Malignant
    Disease, 5
Steiner, Alton, 203
Stern, Joe, 33
Stettin, De Witt, 149
Strittmatter, Philipp, 75
Strominger, Jack, 27, 80
Succinyl CoA, 33
Sucrose, 50, 51
Sulfation factor, 167
Sutherland, Earl W., 50, 53–65, 70, 105, 132, 151,
    154, 156, 203, 217
Sympathetic nervous system, 135, 202
Szent-György, Albert, 4

Teichoic acid, 186
Theorell, Hugo, 24, 104, 114
Thymidine diphosphate (TDP), 182
Thymidine diphosphate aminosugars (TDP
    aminosugars), 182
Thymidine diphosphate glucose (TDP glucose),
    182
Thymidine diphosphate rhamnose (TDP
    rhamnose), 182
Thyroid hormone, 74, 167
Todd, Alexander, Baron of Trumpington, 26, 49
Transmethylation, 94
Treaty of Versailles, 4
Trieste, 1–3
Tuftsin, 80

Ui, Michio, 133
University of Miami, 186
University of Otago, 46
University of Pennsylvania, 97, 140
University of Würzburg, 191–194
Urea, 22, 133
Urey, Harold, 91
Uridine diphosphate acetylglucosamine (UDP
    acetylglucosamine), 50
Uridine diphosphate glucose (UDPG), 27, 49–51,
    152, 182

Uridine diphosphate glucuronic acid (UDP glucuronic acid), 50
Uridyl transferase, 27, 49
U.S. Navy, 83, 171, 206

Vagelos, Roy, 211
Vanderbilt University, 18, 19, 59, 74, 79, 80, 97, 129, 141
Velick, Sidney, 73–76, 139, 172
Viagra (Sildenafil), 136
Villar-Palasi, Carlos, 152, 153
Von Frisch, Karl, 1
Von Liebig, Justus, 12

Warburg, Otto, 7, 18, 22, 104, 212, 216
Washington University, 12, 139, 163, 182, 210
Watson, James, 25
Western Reserve University, 55, 152
Whipple, Alan, 206
Williams, Robert, 166
Wood, Barry, 127, 128, 166
World War I, 7, 100, 127, 149
World War II, 74, 83, 102, 131, 146, 149, 189, 216

Young, Sir Frank, 71, 162

Zamecnik, Paul, 27

www.ingramcontent.com/pod-product-compliance
Ingram Content Group UK Ltd.
Pitfield, Milton Keynes, MK11 3LW, UK
UKHW040003250426
12049UKWH00027B/8